Using
Hoshin Kanri
to Improve the
Value Stream

Using Hoshin Kanri to Improve the Value Stream

Elizabeth A. Cudney

CRC Press
Taylor & Francis Group
Boca Raton London New York

CRC Press is an imprint of the
Taylor & Francis Group, an **informa** business

A PRODUCTIVITY PRESS BOOK

Dedication

This book is dedicated to my father, Dr. John S.W. Fargher, Jr. He showed me the importance of doing what you love and loving what you do. Following in his footsteps as an industrial engineer was the best career decision of my life.

Table of Contents

Acknowledgments

I am indebted to the colleagues with whom I have had the fortune to work and learn from. They have challenged my thinking and showed me what the Toyota Production System is truly about.

I am extremely thankful to Productivity Press for all of their help and for making this a remarkable experience for me—in particular Michael Sinocchi, Senior Acquisitions Editor, and Lara Zoble, Assistant Editor and Project Coordinator. I am also grateful to Ruth Mills for her developmental edits and Marc Johnston for managing the production process.

My greatest appreciation goes to my family for supporting me, believing in me, and their love. Without such support, fulfilling this dream would not have been possible. My husband Brian and our two wonderful children Caroline and Josh bring me the greatest joy.

Introduction

The idea for this book came from seeing how successful companies implemented Lean and how other companies really struggled with making improvements. I have been fortunate to work with companies and organizations that have strategically implemented Lean and Six Sigma. However, the learning stems from working with the companies that cannot bring Lean or Six Sigma to fruition. The key difference has been the strategic vision of the organization.

When a company does not have a strategic vision or leadership does not support the vision, any improvement methodology is bound to fail. It will be viewed as a "flavor of the month" and will only get lip service. By tying improvement methodologies into the strategic plan, the organization becomes aligned to the same goals. Silos are eliminated, and employees work together. The goal of this book is to provide a "how to" to implementing a successful and strategic Lean Six Sigma program.

My objective in writing this book is to illustrate a technique that cascades down the strategic long-term vision of the organization to prioritize the continuous improvement efforts. Past practices use more of a "whack-a-mole" approach to implementing change. If improvements are not tied to the future of an organization, they are in vain. The primary audience for this book is mid- to upper-level managers, Lean champions, Six Sigma Master Black Belts, and Black Belts. The book is not industry-specific and therefore applies to manufacturing, healthcare, government, and service organizations. Ideally, the book is primarily for group learning because a team should determine the strategic goals of the organization.

The biggest decision in writing this book was really how to scope it. Because my goal was to provide a "how-to" implementation, I decided to go into depth with explanations on how to apply specific improvement tools such as 5S, Single-Minute Exchange of Dies (SMED), Standard Work, and Mistake-Proofing. This broadened the scope of the book but added the linkage between the overall strategy of a company using Hoshin Kanri and the actual usage of Lean techniques.

In addition, I invented a manufacturing company, which I have called Carjo Manufacturing, to help illustrate the concepts and drive them home. Carjo Manufacturing is a composite company of my past experiences with various companies. My main decision here was whether or not to make the company a traditional manufacturing company that cuts chips. I strongly believe a process is a process whether it is in manufacturing, healthcare, or service. To keep the example general enough for any audience, I decided to use a traditional manufacturing company, which most people can relate to. In addition, I included a comprehensive glossary of lean terms to aid in understanding the strategic process at the back of the book.

With that being said, it is essential for all organizations to have a long-term strategy and tie their process improvements to this strategy. Although this book uses a manufacturing example, it can certainly be applied in any industry.

Many organizations on the Lean path begin by creating a value stream map. In doing so, a current-state map of how they presently create value for their customers is developed This is then followed by the creation of an enhanced future-state map, incorporating best practices in their processes through research and benchmarking. The final goal of this exercise is to holistically optimize the entire process of value flow by eliminating waste and controlling variation.

However, achieving full implementation of the enhanced future-state value stream map is far more complex than developing it. One of the techniques that companies can adopt to systematically make progress in implementing the envisioned process is Hoshin Kanri (also known as Policy Deployment). This technique encourages employees to reach the root cause of problems before searching for solutions, creating sustainable plans for implementation, incorporating performance metrics, and taking appropriate action for implementation. Although developed in Japan, this technique is based on Deming's classic Plan-Do-Check-Act (PDCA) improvement cycle. Japanese Deming Prizewinners credit Hoshin Kanri as being a key contributor to their business success.

Hoshin Kanri offers an effective way to tie the long-term strategy of the organization to process improvement efforts. Typically, organizations select their kaizen events and process improvement projects on the basis of where they currently feel pain. I refer to this as "whack-a-mole kaizen." If they had a recent rash of external defects, they might decide to initiate a Six Sigma project as a corrective action response to the customer. Significant time and money is involved in running a Six Sigma project and may not be the best tool. In addition, although this is currently where the company is feeling the pain, it may not be the true highest priority project in looking at the big picture.

What is needed is a systems approach that focuses on the long-term vision and strategy of the organization. The time, talents, and money of the organization should focus on improvements that will affect the flow of the entire organization. As such, organizations should consider systems thinking using theory of constraints to ensure a broad impact on the entire organization. This will also greatly increase the momentum of improvement. As improvements are completed, more people throughout the organization will notice their impact. More people will experience the effects quicker, which will drive the participation and involvement of more people.

In addition, Hoshin Kanri cascades the overall strategic vision of the organization throughout all levels, enabling employees to see how they fit into the big picture of the organization. This linkage aligns everyone on the same strategy and vision. By focusing employees in a common direction, the improvements can have a much larger impact in considerably less time. Think about a small team whose members understand what they need to do and how effective they are in working together. Imagine a company of 500 or 1,000 employees all working together to achieve a common goal.

How to Use the CD

Throughout the book, several forms are used, which can be found on the CD included at the back of the book. These templates are designed to help plan and execute your strategic Lean Six Sigma journey. The following section describes the CD contents and user information. I hope you find the templates and tools useful in your strategic deployment.

The CD at the back of the book contains a file, "Strategic Lean Six Sigma," which includes templates and tools for your strategic Lean Six Sigma journey.

The CD template files are as follows:

■ Policy Deployment templates (Word file): This file contains the Hoshin Kanri templates.
■ Red tag (Word file): This file contains the red tag, which should be printed on red paper. You can print these prior to a 5S kaizen event to have ready to tag unnecessary items.
■ 5S worksheet (Excel file): This file contains the 5S worksheet.
■ Standard Work worksheets (Excel file): This file contains the standard worksheets. Each worksheet is labeled on the corresponding tab.
■ Value stream mapping icons (PowerPoint file): This file contains the standard value stream mapping icons.

I hope you enjoy your journey to becoming a strategic-thinking Lean organization.

LEAN AS A COMPETITIVE STRATEGY

1

The first part of this book discusses the need for organizations to embrace and adopt Lean principles for long-term sustainability. In addition, Carjo Manufacturing is introduced to illustrate its reasons for implementing Lean.

Chapter 1

Lean Philosophy as an Enterprise Solution

The United States has lost significant manufacturing jobs. Currently, one-tenth of U.S. workers are employed in manufacturing organizations, which is down from one-third of U.S. workers 50 years ago. This decline stems from a rise in global competition. Companies must cut prices and therefore must find a means for reducing costs to sustain a profit. We need to improve productivity and quality to remain competitive in the global marketplace. The pressing need for increased productivity has driven U.S. industries to an increased awareness of Lean principles. There has been an enormous shift in the manufacturing, healthcare, and service industries with regards to improving processes to address these issues, thereby improving productivity.

Today, irrespective of industry, corporations must focus on speed, efficiency, and customer value to be globally competitive. Lean principles have enabled corporations to achieve significant economic benefits while improving quality, costs, and cycle time. The Lean approach is focused on the identification and elimination of waste. Although Lean principles were originally developed in the automotive industry, they can be applied to any business.

With vast amounts of wasted time and money in many design and manufacturing systems used by modern companies, the need for a way to make processes more efficient and trim is an important step for many corporations. Lean is a philosophy that seeks to increase efficiency and improve the services and products provided by an enterprise. Lean attempts to reduce complication, eliminate wasteful practices, and simplify the business process. Lean can eliminate wasted overhead that drives up operating costs for manufacturers and prices for their consumers. Lean principles are an ongoing consideration that must be maintained and managed to remain effective.

The Benefits of Lean

The purpose of implementing Lean is to improve quality, productivity, profitability, and market competitiveness. Lean focuses on eliminating and preventing waste and improving flow. Lean is focused on the customer by addressing what is value added and what is non-value added. Products and services are delivered just in time (JIT), meaning in the right amounts, at the right time, and in the right condition. Products and services are produced only when a signal is received from the customer and are pulled through the system. A Lean system allows for an efficient response to fluctuating customer demands and requirements.

Lean transformation requires simultaneous changes in the technical system, the behavioral system, and the management system. Implementing the Lean philosophy is a continuing and long-term goal that can deliver some results quickly, but it may take years before the approach becomes a core aspect of an organization's culture.

The seven wastes include:

1. Overproduction
2. Transportation
3. Inventory
4. Overprocessing
5. Waiting
6. Motion
7. Defects

Any activity that consumes resources such as time, manpower, or money, but does not add to the value of the product, is considered waste. These factors can be controlled with careful planning and implementation of Lean methods. To create a Lean system, the value procession must be examined, and the waste or value leaks found. It is important to remember that any wasted resource in the creation of a product means that that product holds less value. In an increasingly competitive business world, providing a service that your customers can afford takes precedence over simply setting prices as "costs plus profit." Maintaining profits and providing affordable service means that costs must be trimmed. Lean can accomplish this while maintaining or even improving quality levels.

Customers continually want more reliable, durable products and services in a timely manner. To remain competitive, all organizations must become more responsive to customers. Lean emphasizes the elimination of waste and creation of flow within an enterprise. The primary focus of Lean is on the customer, to address value-added and non-value-added tasks. Value added tasks are the only operations for which the customer is willing to pay. The idea of creating flow in Lean is to deliver products and services JIT, in the right amounts and at the right quality levels at the right place. This necessitates that products and services are produced and delivered only when a pull is exerted by the customer through a signal in the form of a purchase. A well-designed Lean system allows for an immediate and effective response to fluctuating customer demands and requirements.

An Overview of Lean Tools

Lean tools that are most commonly used to eliminate waste and achieve flow include:

- Value stream mapping
- Standard Work
- 5S housekeeping
- Single-Minute Exchange of Dies (SMED)
- Visual management

Value stream mapping is the first building block in Lean. The purpose of value stream mapping is to understand the big picture. The current value stream consists of all actions necessary to deliver a product including value added and non-value added. Value stream mapping must be conducted first to provide an effective blueprint for implementing an improvement strategy. A key step in creating the current-state map is to calculate takt time.

The Lean philosophy can be used in any area of the business process, be it marketing, manufacturing, design, or even human resources. The range of implementation depends only on the motivation and creativity of the instigating administration. To make the best of Lean principles, the organization must look for new opportunities. Lean as a philosophy is not about just doing better than competitors; it is about going beyond and being the best in every process and product.

What is needed is a strategy to tie together the long-term vision of the organization with the Lean philosophy. This book will show the flow down of the strategic vision through all levels of an organization into the daily management activities. This will enable the proper Lean technique to be used for optimal results that impact the organization as a whole rather than just a business unit.

This book will first introduce you to Carjo Manufacturing, which is a composite company based on my past experiences. Hoshin Kanri, or Policy Deployment, will then be covered in depth.

A Five-Phase Implementation Methodology

The ultimate goal of this book is to illustrate a five-phase methodology of how the implementation of the enhanced future-state value stream map can be expedited using Hoshin Kanri. A graphical representation of the five phases is provided in Figure 1.1. In Phase 1, you start by deploying formalized Lean and variation reduction (or Six Sigma) training. Formal Lean training should include training on the technique followed by an implementation project.

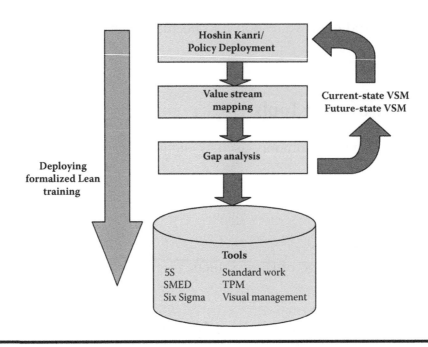

Figure 1.1 Improvement strategy.

At the same time as deploying the formalized training, you begin Phase 2 to capture the strategic goals of the organization. Your goals should then be driven down through the organization and integrated into the daily activities of everyone in the organization.

In Phase 3, value stream mapping is performed to identify all value-added and non-value added steps required to bring a product from raw materials to the customer. Next, you map the current state to identify how the process is currently operating. Kaizen bursts for areas of improvement are then identified. Finally, in Phase 3 you develop the future state to design a Lean flow. Next, in Phase 4 you perform a gap analysis between the current state, future state, and strategic goals to prioritize the identified kaizen bursts. This enables you to properly select the improvement project that will have the greatest impact on the entire organization rather than suboptimal improvements impacting only one area.

In Phase 5 you focus on performing kaizen events. Standard Work and 5S must be top priority as these techniques lay a foundation by improving consistency. Using the prioritized kaizen bursts you develop action plans or schedules to perform the kaizen events or Six Sigma projects.

Conclusion

Lean drives continuous improvement, and as a stand-alone tool it is effective. However, tying Lean initiatives to the strategic vision of the organization considerably enhances the effectiveness of Lean. Chapter 2 introduces you to Carjo Manufacturing and its need to implement Lean.

Chapter 2

Case Study: An Introduction to Carjo Manufacturing Co.

Lean is a philosophy that drives continuous improvement and improves the bottom line for an organization. Now let us take a look at Carjo Manufacturing to see how Lean principles can benefit this organization.

Carjo Manufacturing produces axle components for the automotive industry. The company's main products include shafts, tubes, and gear housing assemblies. The company's sales for 2006 were $18 million but dropped to $16 million in 2007 because of the market, and sales for 2008 are expected to be down again, slightly, from 2007. Therefore, in order to survive, Carjo must implement Lean to drive out wastes and improve its profit margins.

Carjo's Leadership and Management

Management understands the change in the manufacturing environment and sees the need to change in order to survive. In addition, Carjo's senior leadership has heard about the success of Lean and has driven some Lean implementation. The problem has been that the company has seen only modest gains from implementing Lean. The senior leadership team expected more dramatic gains from the kaizen events that had been held. The senior leadership team consists of the following:

- CEO
- Director of Operations
- Director of New Business
- Director of Marketing
- Director of Engineering
- Director of Quality
- Director of Finance

Carjo's Operations at Three Plants

Carjo's operations consist of three plants:

1. The first plant manufactures the shaft for the gear assembly.
2. The second plant machines and assembles the tube.
3. The third plant machines the gear housing and then assembles the gear assembly.

The three facilities are on the East Coast and in close proximity, but in different states. None of the plants are large enough to contain all of the necessary operations as they currently operate.

The production processes for the final product, the gear-housing assembly, involve Plants 1, 2, and 3. Transportation time is 2 days by truck from Plant 1 to Plant 3, 1.5 days by truck from Plant 2 to Plant 3, and 1 day by truck from Plant 3 to the customer.

Various vendors supply the raw materials at each plant. Deliveries of castings, bearings, and bushings occur once a week to Plant 1 for the production of the tubes. Plant 2 receives deliveries of shaft castings every two weeks. Plant 3 machines the gear housing and performs the final assembly. For this process, Plant 3 receives shipments twice a week from Plant 1 (tubes), Plant 2 (shafts), and the gear supplier. Plant 3 then makes daily deliveries to the customer.

The main operations, including Carjo headquarters, are housed at Plant 3. This is where the main production control department resides. Carjo receives its customers' 90/60/30 day forecasts and enters them into MRP. In turn, Carjo issues a 90/60/30 day forecast to all of its suppliers (of tube castings, bearings, bushings, shaft castings, and gears). Carjo secures its raw materials using a weekly faxed order release to its suppliers. Internally, the production control department generates MRP-based, weekly departmental requirements based on customer orders. On the basis of this information, the production control department issues daily build schedules to Plants 1, 2, and 3. Daily shipping schedules are also issued to the shipping departments at Plants 1, 2, and 3.

In terms of the macro level, the process information is shown in Figure 2.1. (Chapter 7 provides more detail in the process-level current-state maps for Plants 1, 2, and 3. The maps are provided in Figures 7.2, 7.3, and 7.4, respectively, for the three plants.)

The macro-level value stream map is shown in Figure 2.2. On the basis of their macro-level value stream map, Carjo's senior leadership team (which consists of the CEO and directors) can see they have a significant problem. Coupled with the current market conditions, Carjo must improve its processes to ensure long-term business viability. Lean has provided some benefits but has not had the overall impact on the business that the leadership team envisioned. There is a disconnect between the improvement activities and where the organization must go strategically.

Plant 1:

- Raw material is shipped in by truck
- Shipments are received once a week, usually on Fridays to ensure the raw material is received prior to the start of the next production week
- Shipments from the casting supplier and bearing supplier take 1 day
- Shipments from the bushing supplier take 2 days
- Lead time is 8.8 days
- No changeover is required
- Process reliability is 84%
- Yield is 80%
- First-pass yield is 63%
- Observed inventory (shown in detail in process-level maps in Figure 7.2):
  - 7,802 tubes (in receiving)
  - 8,986 bushings (in receiving)
  - 6,329 bearings (in receiving)
  - 275 tubes machined and washed (ready for bearing press)
  - 30 final products waiting for packaging
  - 3,870 tubes packaged and waiting for shipment to the customer
- Tubes are shipped to Plant 3 by truck
- Deliveries occur twice per week and take 2 days

Plant 2:

- Raw material is shipped in by truck
- Shipments are received twice a month and take 2 days
- Lead time is 16.2 days
- The plant follows the same process for the two shafts produced
- Changeover is required on most of the processes for the two different types of shafts
- Process reliability is 90%
- Yield is 96%
- First-pass yield is 87%
- Observed inventory:
  - 11,650 shaft castings (in receiving)
  - Various levels of WIP between the processes (shown in detail in process level maps in Figure 7.3)
  - 3,115 shafts packaged and waiting for shipments to the customer
- Finished shafts are shipped to plant 3 by truck
- Deliveries occur twice per week and take 1½ days

Plant 3:

- All material is shipped in by truck (from Plant 1, Plant 2, and gear supplier)
- Shipments are received twice a week

Figure 2.1   Process information flow at Carjo Manufacturing Co.

■ Shipments from Plant 1 and the gear supplier take 2 days
■ Shipments from Plant 2 take 1½ days
■ Lead time is 9.4 days
■ No changeover is required
■ Process reliability is 95%
■ Yield is 88%
■ First-pass yield is 80%
■ Observed inventory:
  ○ 4,492 shafts (in receiving)
  ○ 2,783 shafts (in receiving)
  ○ 2,314 gears (in receiving)
  ○ Various levels of WIP between the processes (shown in detail in process level maps Figure 7.4)
  ○ 903 gear-housing assemblies, packaged and waiting for shipments to the customer
■ Finished assemblies are shipped to the customer by truck
■ Deliveries occur daily and take 1 day

Figure 2.1   Process information flow at Carjo Manufacturing Co. (Continued).

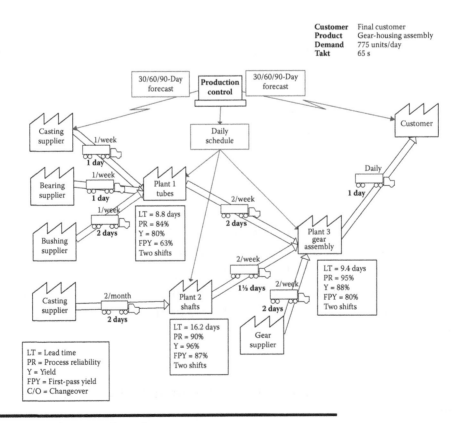

Figure 2.2   Carjo Manufacturing Co. macro current-state map.

Conclusion

Carjo has realized modest gains from implementing Lean techniques. However, Carjo can achieve more significant results by tying its improvement efforts to its strategic long-term objectives. The next part of the book takes you through strategic thinking by introducing Policy Deployment in Chapter 3 and Hoshin methods in Chapter 4. Chapter 5 then shows you how Policy Deployment is implemented at Carjo Manufacturing.

STRATEGIC THINKING

Now that you understand the need to implement Lean, the next step is to link Lean initiatives to the strategic goals of the organization. Part II covers Hoshin Kanri in detail to show you how to implement strategic thinking into your organization. Carjo's implementation of Hoshin Kanri is also discussed in detail to provide you with an illustrative example to further drive home the concepts.

Chapter 3

Integrating Strategic Goals

The Lean philosophy drives continuous improvement; however, to realize significant improvements you must link Lean efforts to the strategic vision and goals of your organization. This will ensure the most appropriate projects are implemented to achieve the greatest gain for your organization. The goal of this chapter is to begin linking the strategic goals with your business system. Later in Chapter 16 you will learn how to integrate your strategic goals into your organization's daily activities.

The Japanese quality thinking began before 1645. Miyamoto Musashi wrote a guide to samurai warriors on strategy, tactics, and philosophy entitled *A Book of Five Rings*. Musashi was a Japanese swordsman who became legendary for his duels and distinctive style of swordsmanship. In his book, Musashi states, "If you are thoroughly conversant with strategy, you will recognize the enemy's intentions and have opportunities to win."

A corporation's strategic plan must be integrated with the macro-level value stream map to identify the optimal improvement opportunities. This promotes strategic thinking. Often improvement activities are identified with silo thinking. The effects on other systems or processes within the organization are not considered. Improvements in one area can have a negative impact on another business area.

Senior leadership, including the CEO and directors, should use Hoshin Kanri to develop long-term strategic objectives. Mid-level managers should then use macro-level value-stream mapping to identify areas of improvement to achieve strategic goals. Finally, department teams should use Lean and Six Sigma tools for process improvement.

To think strategically, the senior leaders of a corporation—in other words, you—should first determine where it is going—the vision. Then you need to identify your business' key processes. Next, you should perform a gap analysis between your organization's current state and your vision. This will lead to a strategic approach to continuous improvement.

To become a Lean enterprise, you must integrate Lean throughout all levels of your organization. This means breaking down the silos and changing the focus of process improvement to a global perspective. What you need is a holistic approach to continuous improvement throughout your corporation. This will enable your corporation to make improvements that get more significant bottom line results, rather than suboptimal improvements. Lean strategy deployment breaks down these barriers and enables a holistic approach to continuous improvement that links to the long-term goals of your corporation.

A Brief History and Description of Policy Deployment

Dwight D. Eisenhower was quoted as stating, "Plans are nothing, planning is everything."

Hoshin Kanri began in Japan in the early 1960s as statistical process control (SPC) became total quality control (TQC). Hoshin Kanri is most commonly referred to as Policy Deployment (PD). "Hoshin" means shining metal, compass, or pointing the direction. "Kanri" means management or control. Here is an overview of what PD is and does.

- PD is a systems approach to management of change in critical business processes.
- It is a methodology to improve the performance of critical business processes to achieve strategic objectives.
- PD improves focus, linkage, accountability, buy-in, communication, and involvement in a corporation.
- It links business goals to the entire organization, promotes breakthrough thinking, and focuses on processes (rather than tasks).
- PD is also a disciplined process that starts with the vision of the organization to develop a 3- to 5-year business plan and then drives down to one-year objectives that are deployed to all business units for implementation and regular process review. PD is a business management system designed to achieve world-class excellence in customer satisfaction. The system, beginning with the voice of the customer, continuously strives to improve quality, delivery, and cost. The system provides the necessary tools to achieve specific business objectives with the involvement of all employees.

As shown in Figure 3.1, you should take the voice of the customer to drive your business targets. Then, using PD as your management strategy, you should drive down this strategy throughout all levels of your business to focus on safety, quality, delivery, and cost. Then, using foundational Lean Six Sigma tools such as pull, 5S, Single-Minute Exchange of Dies (SMED), Standard Work, Mistake-Proofing, and value stream mapping, you can focus on continuous improvement. This leads to improved customer satisfaction, which further leads to improved sales growth for your organization.

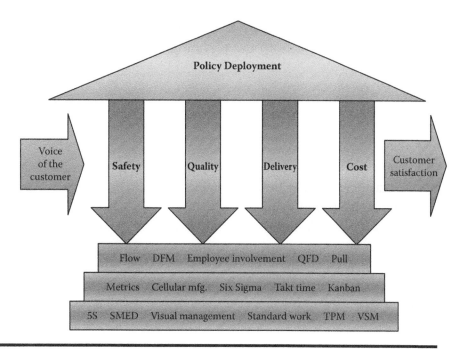

Figure 3.1   Strategic business system.

Two Levels of PD

PD is a methodology to capture strategic goals and integrate these goals with your entire organization's daily activities. The two levels of PD include: (1) management or strategic planning and (2) daily management.

Planning for PD

Effective planning is critical for the long-term success of a corporation. There are five main steps for effective planning:

1. Identify your critical objectives.
2. Evaluate the constraints.
3. Establish performance measures.
4. Develop an implementation plan.
5. Conduct regular reviews.

Daily Management of PD

Daily management involves applying Plan-Do-Check-Act (PDCA) to daily, incremental, continuous improvement to identify broad system problems in your organization. Once you gain a breakthrough improvement in the system problem, then the improvement becomes the focus of daily continuous improvement activities. Hoshin planning is the system that drives the continuous improvement and breakthroughs. PD involves both the planning and deployment.

■ Develop your targets.
■ Develop your action plans to achieve your targets.
■ Deploy both.

The concept of hierarchy of needs was introduced by Maslow and outlines the basic needs that must be met before moving on to a higher need. Maslow's hierarchy of needs is illustrated in Figure 3.2.

In concurrence with the hierarchy of needs that must be met for an organization to move on to its higher need, there are five levels of organizational needs:

1. Core vision
2. Alignment
3. Self-diagnosis
4. Process management
5. Target focus

Figure 3.3 illustrates the five levels of organizational needs.

The five levels of organization needs are directly linked to the six Hoshin planning steps and five Hoshin methods. Figure 3.4 shows the linkage between the three.

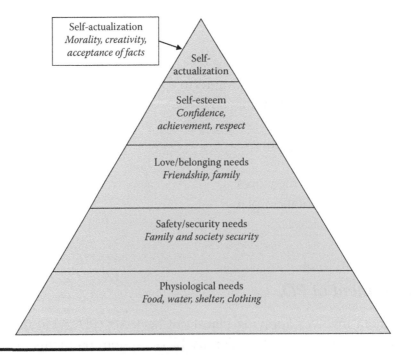

Figure 3.2 Maslow's hierarchy of needs.

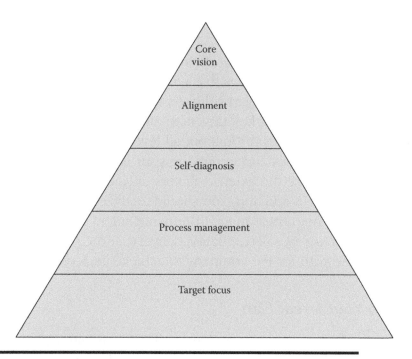

Figure 3.3 Levels of organizational needs.

Within the Hoshin planning process, there are six main steps. Each of these steps is discussed in the next paragraphs.

Step 1: Develop a 5-Year Vision

Top management should develop a 5-year vision to define the strategic objectives of your organization on the basis of the internal, external, and environmental challenges your organization faces. The 5-year vision is the future target for your organization. It is defined by your organization's goals, capabilities, and culture. It is a statement of where your organization wants to be in the future. The vision is a communication tool for senior leadership to relate the ideal future for the organization. For example, Company XYZ has a vision to be a top-10 company in its field in 5 years.

Organizational needs	Hoshin planning steps	Hoshin methods
Core vision	5-Year vision 1-Year plan	Hoshin strategic plan summary
Alignment	Deployment	Hoshin plan summary
Self-diagnosis	Implementation	Hoshin action plan
Process management	Monthly reviews	Hoshin implementation plan
Target focus	Annual review	Hoshin implementation review

Figure 3.4 Linkage of organizational needs, Hoshin planning steps, and Hoshin methods.

Step 2: Develop a 1-Year Plan

On the basis of your 5-year vision, you should then develop a 1-year plan to outline continuous improvement activities that will enable your company to achieve its long-term strategy. The 1-year plan is linked to the 5-year vision by taking an incremental step and defining key targets to attain that step. The purpose of this step is to focus the activities throughout all levels of your organization on addressing your external issues and improving your internal problems. As part of this step, you should analyze the external factors, including your competition and the economy as a whole. In addition, you should analyze past problems so that your organization does not repeat them. Top management must then prioritize the objectives on the basis of safety, quality, delivery, and cost. Developing the 1-year plan blazes the path for the company to achieve its 5-year vision.

Step 3: Deploy Your 1-Year Plan

The next step is to deploy your 1-year plan to all departments within your organization. Deploying your 1-year plan is where you begin to set measureable goals for each department. This is a planning step to determine specific opportunities for improvements within each department. At this point, a strategy is set to achieve the path set by the 1-year plan. This leads to Step 4, which is implementation.

Step 4: Implement Continuous Improvement Activities

Each department must drive continuous improvement activities that are aligned with your 1-year plan and 5-year vision. This step is where the improvement process begins. The prior steps (1–3) involved planning for improvement; now you are performing the improvement activities. For example, an improvement activity in this step could be a SMED event in a cell to reduce changeover time. The improvement activities in Step 4 must tie directly back to the 1-year plan. This involves developing a master plan with appropriate measures and goals.

Step 5: Conduct Monthly Reviews

You should track the progress of your continuous improvement activities using quantitative metrics and communicate them to senior leadership (i.e., the CEO and directors) in a monthly review. The monthly reviews should link directly to your deployment of the 1-year plan. You should monitor your actual improvements against your planned improvements as a monthly self-diagnosis and to ensure the corrective actions are sustained. During a formal review to management, each department should present the tasks addressed, a problem analysis, and the problem-solving results. As the Hoshin planning implementation becomes more engrained into your organization, the review may become more of a highlight or overview of the problems and corrective actions.

Step 6: Perform an Annual Review

Finally, you should conduct an annual review to monitor your progress and capture your organization's results. The annual review provides an opportunity to ensure that the implemented projects helped attain the 1-year plan and 5-year vision. This is a check of how your implementation affected the organizational metrics set out in your 5-year vision. On the basis of the results of the previous year and the effectiveness of your implementations, a new 1-year plan is developed to set the targets and goals for the upcoming year. In addition, the organization may redevelop its 5-year vision on the basis of the current business environment.

Conclusion

The strategic planning process is a critical piece of the puzzle because it lays the foundation for the next step when you will cascade your organization's strategic goals throughout your organization. To accomplish the six Hoshin phases, there are Hoshin Methods that align with the phases; the next chapter discusses 4 of these methods in detail (and 2 more are discussed in Chapter 16).

Chapter 4

Hoshin Methods

As you learned in Chapter 3, effective planning is critical for creating an organizational strategy and vision. The next step is to cascade your strategic vision and goals down into all levels of the organization. This chapter discusses the first four of the six Hoshin methods:

1. Hoshin strategic plan summary
2. Hoshin action plan
3. Hoshin implementation plan
4. Hoshin implementation review

Two additional Hoshin methods are discussed later in Chapter 16 to drive daily management after the process improvement tools are introduced.

Develop a Hoshin Strategic Plan Summary

The first step in Policy Deployment is to develop your strategic plan summary. The Hoshin strategic plan summary links the strategic vision of the organization with measurable goals. The Hoshin strategic plan summary shown in Figure 4.1 illustrates the relationship between an organization's strategic goals, core objectives, metrics, and ownership. A value stream map gives you a picture of all of the activities required to produce a product (Chapter 6 discusses the basics of value stream mapping). The Hoshin strategic plan summary also provides a picture of the overall strategy of an organization and how the strategy cascades throughout all levels of the organization. The linkage is clear on how each strategic goal is measured and who has the ultimate responsibility.

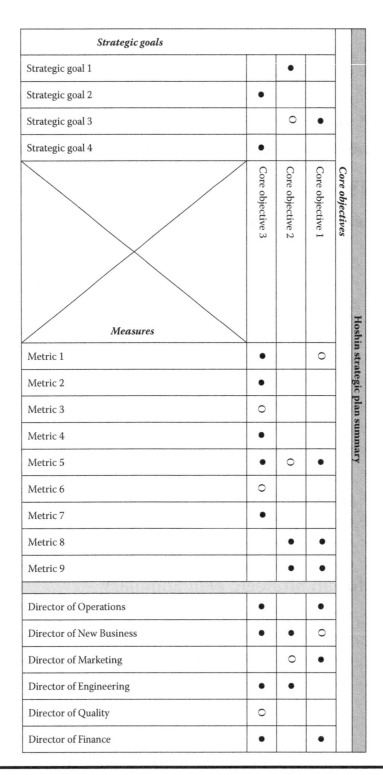

Figure 4.1   Carjo's Hoshin strategic plan summary.

How to Create a Hoshin Strategic Plan Summary for Your Organization

Let us look at each aspect of Figure 4.1 to illustrate how you can develop your own Hoshin strategic plan summary. (Figure 5.1 shows what a completed summary looks like for the Carjo Manufacturing Co.; Chapter 5 describes it in detail.)

List Your Strategic Goals

In Figure 4.1, the strategic vision is listed vertically on the left-hand side under the heading "Strategic Goals." These are the broad strategic goals of the organization for the next five to ten years. The strategic goals should be what the organization needs to do to ensure long-term success.

List Your Core Objectives

Next, we drive down a level to get more specific measurable goals. As you can see in Figure 4.1, these specific measurable long-term goals (for the next three to five years) are listed horizontally under the heading "Core Objectives."

Make Sure Your Core Objectives Link to Your Strategic Goals

To ensure these goals are linked to the overall strategic goals in the plan summary (in Figure 4.1), a coding system is used to show the strength of the linkage:

■ A filled-in circle (•) indicates *a strong relationship* between the strategic goal and the core objective.
■ An open circle (◦) indicates *a direct relationship* between the strategic goal and the core objective, but the core objective is not necessarily one of the key drivers for that strategic goal.

The key reason for showing the relationships in the strategic plan summary is to ensure that you are adequately addressing the strategic goals of your organization through your core objectives and that you are adequately measuring them by using the appropriate metrics.

Identify Who Is Responsible: Identify Your Metrics

Finally, you also want to ensure that the proper ownership exists to drive the necessary improvements. In other words, you need to assign each improvement activity to a specific person for accountability.

Determine How You Will Measure Improvements for Meeting Each Goal

Now, you cascade the overall objectives into how you will manage the business and measure the progress of your process improvement activities. In this step, you now need to determine your metrics and short-term goals (for the next one to two years). These metrics will drive how you will manage your business and how you will prioritize your process improvement activities. List your metrics vertically on the right side of the matrix, as shown in Figure 4.1.

Make sure you tie your metrics to your core objectives. To ensure your organization will meet your measurable core objectives, these metrics must meet several criteria.

■ First, they must be *measurable* (quantitative not qualitative). These metrics assess the effectiveness of your process improvement efforts.
■ The metrics must also be *baselined* to show your current performance or *benchmarked* against your competitors or your industry's standards.
■ In addition, these metrics must be *achievable*. When your metrics are unachievable (i.e., zero internal defects), your employees will become discouraged, and your system will fail. In contrast, if you set realistic goals (i.e., zero external defects), your employees will team together to ensure your success.

Next, use the same coding scheme.

■ A filled-in circle (•) to show strong relationships between your metrics and your core objectives.
■ An open circle (◦) to show direct relationships between your metrics and your core objectives.

Identify Who Is Responsible for Meeting Your Core Objectives: Assign Ownership

The final step in the strategic plan summary is to assign ownership of the core objectives. The ownership at this level falls on your organization's executive leadership because they own the responsibility of creating the strategic vision and driving it down through your organization. In Figure 4.1, the leadership team is listed vertically at the far right of the matrix. Here, again, we use the coding scheme to illustrate ownership.

■ Show ownership of a core objective with a filled-in circle (•).
■ Show cursory responsibility with an open circle (◦).

Because these are tied into the strategic goals of your organization, there will be core objectives with several owners. For example, if one of your strategic goals is to drive new business through your product offerings, this will involve engineering and marketing (as well as possibly others). Therefore, in this example, both engineering and marketing would have the main ownership.

Drive Your Strategy Down to the Department Level—Develop a Hoshin Plan Summary

The Hoshin plan summary details the strategic goals and cascades them down to the department level. Whereas the Hoshin strategic plan summary was at the highest organization level, the Hoshin plan summary is the tactical plan for each department. Now you need to drive the strategy down to each department. Figure 4.2 is a blank template of the Hoshin plan summary. (Figure 5.2 shows a sample Hoshin plan summary for Carjo Manufacturing.)

The first column in Figure 4.2, labeled "Strategic Goals," should correspond to the strategic goals listed in your Hoshin strategic plan summary (in Figure 4.1). Now that you are moving down to the department level, the management owner will be the person in charge of that department. Each department will have its own Hoshin plan summary. In some cases, depending on the management structure of your organization, the management owner will be the same here as the core objective owner in your Hoshin strategic plan summary.

The next two columns are for your short- and long-term goals. Your long-term goals may, in some cases, correspond to the strategic goals from the left side of your Hoshin strategic plan summary. However, your goals should definitely correspond to the metrics you outlined in the Hoshin strategic plan summary. This ensures that you align the proper activities with your overall strategic vision. On the basis of the metrics you previously outlined, your short- and long-term goals should already be developed. These may not necessarily be the same for each department.

For example, let us look at external defects (measured as external parts per million [PPM]): a manufacturing department's short- and long-term goals should be very aggressive. On the other hand, the engineering department should also be focusing on improving the product design using tools, such as design for manufacture and assembly, which will in turn reduce external defects. Therefore, engineering department's short- and long-term goals for reducing external defects will not be as aggressive. Also, your organization as a whole must be aligned and managed by your senior leadership team to ensure that your various departments come together to provide the overall necessary reduction in external defects that your organization is seeking. This provides a common goal for multiple departments to work together to achieve and eliminates silos.

In addition, with respect to the goals, there may be several metrics that relate to a core objective. As noted in Figure 4.2, you may need to list your core objective in multiple rows to correspond to the appropriate metrics. The various metrics for a core objective may then call for different implementation strategies. For example, if your core objective is to improve product quality, this can be measured with internal PPM and external PPM. Internal PPM may be handled with an internal Six Sigma project as the implementation strategy. On the other hand, the implementation strategy to reduce external PPM may be to implement a poke-yoke device. So, you would want to list these in two different rows to highlight that they are two different metrics with different implementation strategies.

Hoshin plan summary

Strategic goals	Management owner	Goals		Implementation strategies	Improvement focus			
		Short-term	Long-term		Safety	Quality	Delivery	Cost
Strategic Goal 1								
Strategic Goal 1								
Strategic Goal 2								
Strategic Goal 2								
Strategic Goal 2								
Strategic Goal 3								
Strategic Goal 4								
Strategic Goal 4								

Figure 4.2 Hoshin plan summary.

Develop Implementation Strategies for Your Hoshin Plan Summary

The next step in developing your Hoshin plan summary is to develop your implementation strategies. This is critical in how your organization makes process improvements appropriately, using the most efficient and effective technique. Each department must develop a strategy on how it will achieve its short- and long-term goals. The team members developing the strategy should revisit their current-state map(s) to understand all of the activities involved. This will enable them to select the most effective technique—whether it is a Six Sigma project, 5S, Standard Work, Single-Minute Exchange of Dies (SMED), Mistake-Proofing, etc.

Decide Where You Want to Focus Your Improvement Efforts

The final step in completing your Hoshin plan summary is to determine your improvement focus. Typical focus areas for organizations include safety, quality, delivery, and cost. Here, again, use the same coding scheme:

- Use a filled-in circle (•) to show strong relationships between the implementation strategy and its impact on safety, quality, delivery, and cost.
- Use the open circle (○) to show cursory relationships.

The purpose of showing the relationships in this matrix is slightly different. Here, you want to balance your improvement efforts. You still need to ensure that you link your implementation strategies to your improvement goals, which are linked to your core objectives. This common thread of linkage must be clear. But you also want to make sure that the implementation strategies you develop will have an impact on your improvement focus areas. For example, if you develop an implementation strategy that only impacts quality but does not impact safety, delivery, or cost, that may be a signal that it is not the most effective strategy. You want an implementation strategy that impacts more than one focus area. As with anything, there must be a balance. If you have a critical safety issue, then it should probably take precedence and may not impact any of the other improvement focus areas. But, in general, because you will be expending time and money for process improvement, you would want to impact multiple focus areas.

Develop Your Hoshin Action Plan

Next, you need to develop your Hoshin action plan. This further drives down your core objectives into the daily activities of your organization for process improvement by creating a detailed action plan. You should present this action plan to your leadership at a set frequency (typically a weekly walkthrough and/or monthly management review).

Figure 4.3 shows a blank template of a Hoshin action plan. The top portion of the Hoshin action plan provides the necessary information to show the linkage between your action plan and each of your strategic core objectives. The following information is necessary:

- Core objective
- Management owner
- Department
- Team
- Date
- Next review

To illustrate the Hoshin action plan, let us continue on with the example of improving product quality.

The next section is the situation summary. The situation summary provides a problem statement of the current status. It should clearly state why the improvement is necessary. An appropriate situation summary might be:

> Product quality is a key market driver in our industry. The external PPM for Product A has increased from 3,861 PPM to 4,725 over the past six months. As a result, our supplier rating has dropped from an A in the first quarter to a B in the second quarter.

Hoshin action plan	
Core objective: Management owner: Department:	Team: Date: Next review:
Situation summary:	
Objective:	

Short-term goal: Long-term goal:	Strategy:	Targets and milestones:

Figure 4.3 Hoshin action plan.

Next, define your overall objective. This objective should relate back to one of your core objectives. The objective statement might then be "to improve product quality by decreasing external defects by 50%."

The next step is to complete the short- and long-term goals using the metrics you previously detailed in your Hoshin strategic plan summary (shown in Figure 4.1). You should identify your short-term goals for a period of the next three to six months, and your long-term goal should focus on improvements for the next twelve months.

The next step is to discuss the implementation strategy. This should flow down from your Hoshin plan summary (shown in Figure 4.2); however, at this point it should be more detailed. For example, with the core objective of "improving product quality," your implementation strategy in the Hoshin plan summary might have simply stated, "Six Sigma Project." In the Hoshin action plan, you would want to clarify this in more detail; for example, you might explain this as "Six Sigma Project on oversize cylinder bore."

The final step in the Hoshin action plan is to outline the targets and milestones of your strategy. In continuing with the example of the Six Sigma project on the oversize cylinder bore, your targets might be "perform a hypothesis test" or "run a design of experiments." The milestone would be the anticipated completion date. (Refer to Figure 16.1 for an example.)

Develop Your Hoshin Implementation Plan

Now you should develop your Hoshin implementation plan, which records your progress and lists the implementation activities. Figure 4.4 is a template you can use for this. Your implementation plan compares the current status of milestones to your initial projections. It is typically shown in a Gantt chart format.

Review your Hoshin implementation plan with your organization's senior leadership at a set frequency, typically monthly. This requires each department to outline its expected improvement gains by month for the following year.

The top portion of the Hoshin implementation plan details each core objective, its management owner, and the date targeted to achieve that objective. Your core objectives on the Hoshin implementation plan should link back to the high-level Hoshin strategy summary plan (shown in Figure 4.1), and the management owner should link back to the Hoshin plan summary (shown in Figure 4.2).

In the first column on the left of Figure 4.4, list your implementation strategies, as outlined by each department. Each department should have a Hoshin action plan for each implementation strategy.

In the next column of Figure 4.4, define your target and the actual performance for each implementation strategy. The performance should be measured

		Hoshin implementation plan												
Core objective:														
Management owner:														
Date:														
Strategy	**Performance**	**Schedule and milestones**												
		Jan	Feb	Mar	Apr	May	June	July	Aug	Sept	Oct	Nov	Dec
	Target												
	Actual												
	Target												
	Actual												
	Target												
	Actual												
	Target												
	Actual												
	Target												
	Actual												

Figure 4.4 Hoshin implementation plan.

using the metrics defined in your Hoshin strategic plan summary (Figure 4.1). Then, break down the performance by month to monitor the performance improvement trends for the year. One way to visually show which metrics are on track by month is to color the background of that month's performance in green; months that did not meet the target performance can then be colored in red. This makes it easy for you to hone in on those implementation strategies that are not meeting their target performance.

Conduct a Hoshin Implementation Review

Finally, you should conduct a Hoshin implementation review, which records the progress of your performance. Figure 4.5 provides a blank template. The Hoshin implementation review also records your company's performance relative to your industry's overall performance. The implementation plan also lists your highest priority implementation issues.

During your presentation to your senior leadership team, use your Hoshin action plans as backup information to show what targets and milestones you have met as well as a recovery plan to get performance back on track.

There are five key steps for implementing an effective strategy. These are described in the next sections.

Hoshin implementation review	
Core objective:	
Strategy owner:	
Date:	
Performance status	**Implementation issues**

Figure 4.5 Hoshin implementation review.

Step 1: Measure Your Organization's System Performance

In measuring your organization's system performance, it is critical to develop a plan to manage the strategic-change objectives. The initial direction must be adaptable. The planning process must also be adaptive to respond to business changes. Then, regular assessments of planning and implementation are necessary.

Step 2: Set Your Core Business Objectives

To set your core business objectives, a technique called "catchball" is effective to incorporate group dialogue. Catchball is equivalent to tossing an idea around, which provides the optimal objectives for the overall business system.

Step 3: Evaluate Your Business Environment

The business environment must then be evaluated to understand the needs of the organization's customers. These customers include stockholders, employees, external customers, etc. The environmental analysis includes the technical, economic, social, and political aspects of the business. The purpose is to answer the question "How does the business perform relative to their competitors?"

Step 4: Provide the Necessary Resources

For the strategic alignment to be successful, management must also provide the necessary resources to lead the efforts for both the strategic objectives and daily management. Remember—the purpose of Hoshin is to align the system to strategic-change initiatives. This requires resource commitment.

Step 5: Define Your System Processes

Another key aspect is to define the system processes. Hoshin enables consensus planning and execution between all levels of the organization, as shown in Figure 4.6. The Hoshin plan aligns the strategic vision, strategy, and actions of the organization. The actions of senior management, middle management, and the implementation teams (all levels of the organization) are aligned around the common Hoshin plan. The next section of this chapter discusses the three main tools of Policy Deployment for gaining consensus.

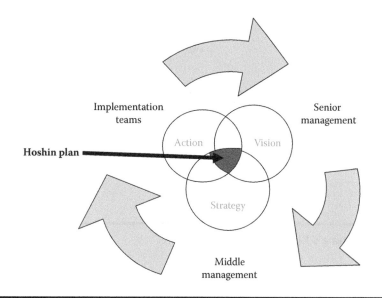

Figure 4.6   Hoshin plan alignment.

The Three Main Tools of Policy Deployment

Policy Deployment is one of the pillars of total quality management (TQM). TQM is based around Deming's Plan-Do-Check-Act (PDCA) cycle. The three main tools of Policy Deployment are as follows:

1. PDCA cycle
2. Cross-functional management
3. Catchball

Let us look at each one in more detail to show how you can use these three tools for consensus planning and execution.

Deming's PDCA Cycle

Deming developed the Plan-Do-Check-Act cycle (shown in Figure 4.7) as an iterative four-step problem-solving process.

1. "Plan" consists of establishing objectives and processes to achieve specific results.
2. The "Do" step involves implementing the processes.
3. In "Check," the processes are monitored and evaluated against the specifications.

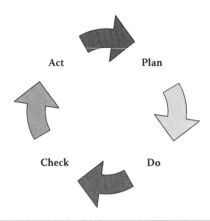

Figure 4.7   Deming's Plan-Do-Check-Act cycle.

4. In the fourth step, "Act," actions are taken to improve the outcome to meet or exceed the specifications.

One of the key differences between PDCA and Hoshin (a.k.a. policy deployment) is that policy deployment *begins* with the check step, which is Step 3 of Deming's cycle. Therefore, the cycle is really CAPD—check-act-plan-do. You start by checking the current status. This propels the Hoshin process. Each company-wide check begins the deployment of a new target and an action plan for achievement.

Cross-Functional Management

Cross-functional management (CFM) enables the continuous checking of targets and means throughout the product development and production processes. It is critical when developing the Hoshin plan to include a cross-functional group from marketing, design engineering, quality, manufacturing, finance, operations, and sales depending on your organization's structure. This enables the diverse group to address the needs of all of the shareholders (internal customers, external customer, stockholders, etc.) in the Hoshin plan. By involving a cross-functional group, you can ensure a balanced representation of your organization's customers' needs.

Catchball

Catchball involves continuous communication. It is essential for the development of targets and action plans. Catchball is also essential for deployment throughout the organization. You must create feedback systems that allow bottom-up, top-down, horizontal, and multidirectional communication. There must be commitment to total employee involvement.

Conclusion

Now that you understand how to implement the four Hoshin methods, let us take a look at how Carjo Manufacturing implemented the four Hoshin methods into its organization.

Chapter 5

Case Study: Implementing Policy Deployment at Carjo Manufacturing Co.

Chapter 2 discussed Carjo's current situation and why it needed to implement Lean principles. Chapters 3 and 4 discussed Policy Deployment and the four Hoshin methods. Now let us take a look at how Carjo implemented Policy Deployment.

The leadership team at Carjo has identified a disconnect between the long-term strategy of the business and the improvement efforts that have taken place. The company leaders have decided to use Hoshin Kanri to flow down the long-term vision of the organization and use it as the strategy for business process excellence.

Developing Carjo's Hoshin Strategic Plan Summary

The senior leadership team at Carjo Manufacturing scheduled a leadership retreat to develop the long-term vision of the organization. The team met for 3 days and developed the following four strategic goals for the organization:

1. Develop world-class new business/marketing.
2. Expand product offerings.
3. Implement cost-reduction projects
4. Implement kaizen projects.

The senior leadership team then used the strategic goals to develop three core objectives for the organization:

1. Improve financial returns by 5% by 2012.
2. Grow sales by $3 million by 2012.
3. Achieve world-class supplier status by 2012.

The leaders then reviewed the relationships between their strategic goals and their core objectives to ensure all of their strategic goals were being addressed properly. Next, they identified the appropriate metrics that would tie directly to their core objectives. The metrics needed to be quantitative (and not qualitative) to indicate whether the process improvements have an impact on the overall organization and are trending in the right direction. Finally, the senior leaders assigned ownership of the core objectives to specific member(s) of their team, as shown in Figure 5.1 (and refer back to Figure 4.1, which provides a blank template).

Developing Carjo's Hoshin Plan Summary

The leadership team next focuses on developing the Hoshin plan summary (see Figure 5.2 and refer back to Figure 4.2, which provides a blank template). As you can see, the strategic goals and owners have been carried down from the company's Hoshin strategic plan summary (refer back to Figure 5.1). The short- and long-term goals must again tie back to the measures outlined in the Hoshin strategic plan summary. The leaders at Carjo decided to have their short-term goals focus on improvements in the next year and the long-term goals three years out.

The next big decision for Carjo's leaders was to determine what their implementation strategies would be based on their strategic goals. For example, in order to develop world-class new business/marketing, the leaders determined their implementation strategy would be to grow sales. Then, on the basis of the implementation strategy, the leadership team determined which improvement focus area was impacted. Using the four improvement focus areas, the leadership team at Carjo outlined the impact of each strategic goal (again, see Figure 5.2). On the basis of this information, the leadership team could clearly see that two strategic goals impacted all four improvement focus areas. These two strategic goals are: (1) implement cost-reduction projects, and (2) implement kaizen projects because they show a relationship with each improvement focus.

This practice helped Carjo's senior leaders prioritize their implementation strategies. Because two of the implementation strategies have a significant impact, these should be the company's highest priority.

Conclusion

Now that Carjo has developed its Hoshin strategic plan summary, its Hoshin plan summary, and its implementation strategies, each department should utilize

Hoshin strategic plan summary

Core objectives
- Improve financial returns by 5% by 2012
- Grow sales by $3M by 2012
- Achieve world-class supplier status by 2012

Measures
- Reduce cost of poor quality (COPQ) by $5.67
- External quality to 4.9σ
- Internal quality to 4.1σ
- Kaizen event savings of $1.2M
- 100% Kaizen participation
- Material productivity to 10%
- Labor and overhead productivity to 10%
- New sales growth to $1M by 2010
- New sales projects of $2M for 2010

Strategic goals
- Develop world-class new business/marketing
- Expand product offerings
- Implement cost reduction projects
- Implement kaizen projects

Resources:
- Director of Operations
- Director of New Business
- Director of Marketing
- Director of Engineering
- Director of Quality
- Director of Finance

Figure 5.1 Carjo's Hoshin strategic plan summary.

Hoshin plan summary								
Strategic goals	Mgt. owner	Goals		Implementation strategies	Improvement focus			
		Short-term	Long-term		Safety	Quality	Delivery	Cost
Develop world-class new business/ marketing	Director of New Business	$1M by 2010	$3M by 2012	Grow sales				●
Expand product offerings	Director of Engineering	$2M for 2010	$5M for 2012	New sales projects		O		●
Implement cost reduction projects	Director of Operations	$1.2M annual savings	$4M annual savings	Improve financial returns	O	●	●	●
Implement kaizen projects	Director of Operations	100% participation	100% participation	Achieve world-class supplier status	●	●	O	●

Figure 5.2   Carjo's Hoshin plan summary.

current-state value stream mapping to identify improvement opportunities relating to the implementation strategies developed to this point. Chapter 6 takes you through value stream mapping, and then we move to identifying improvement opportunities at Carjo Manufacturing in Chapter 8.

UNDERSTANDING VALUE

Now that you understand the strategic goals of your organization, it is important for you to identify and eliminate waste in your processes. Part III takes you through value stream mapping to identify improvement opportunities.

Chapter 6

The Basics of
Value Stream Mapping

Now that your organization has determined its strategic goals, you need to base-line the current status of the critical processes. Current-state value stream mapping is a valuable tool for understanding and documenting an entire process. This chapter provides a quick review of value stream mapping so that you can see how value flows within your processes.

The Purpose of Value Stream Mapping

Value stream mapping (VSM) is a flowcharting method originally created in the Toyota Production System (TPS) to document the entire process (of a company or a department) on a single sheet of paper to encourage dialogue and under-stand the process better. First, you create a current-state map of how the value presently flows in your organization. Then, utilizing the principles of Lean, you envision a better state for how value *should* flow in an optimum manner in your organization: this is *future-state VSM*. Kaizen activities, which are events to overcome deficiencies in the current state that will allow the company to reach the future state, are identified for implementation on the current-state map. The future-state map illustrates the ideal state after the changes are implemented. The future-state map represents how your processes *should* flow.

When you use VSM, you can simultaneously analyze the flow of information and material to enable your organization to eliminate waste in both. By consid-ering material flow and information flow concurrently, it is easy to identify and correct whether material flow is hampering information flow, or vice versa. The overall steps of VSM are shown in Figure 6.1.

VSM captures the flow of a product from the point that raw material enters the process to the point where a final product is delivered to the customer. This

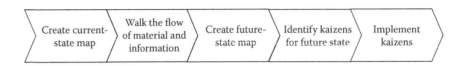

Figure 6.1   Value stream mapping steps.

includes all value-added and non-value-added activities that have been accepted as necessary to produce the product. You can apply VSM to any product or service; therefore, it is finding widespread application to non-manufacturing processes, such as designing a product and to business processes such as purchasing, billing, and selling.

Also, because VSM enables visualization of an entire process as a holistic set of operations rather than as a disjointed movement of material through discrete operation steps, it enables interaction between the operators to help see how they can optimize an entire process, rather than seeking out only local optimization for specific operations. In doing so, the seven classical wastes (as follows) become evident:

1. Overproduction
2. Transportation
3. Inventory
4. Overprocessing
5. Waiting
6. Motion
7. Defects

VSM comprises all of the value-added and non-value-added activities required to manufacture a product. The essential process flow can include bringing a product from raw materials through delivery to the customer or designing a product from concept to job one production. A *value-adding activity* is any activity that transforms raw material to meet your customers' requirements. *Non-value-adding activities* are those activities that take time, resources, or space but do not add to the value of your product or service. The customer is not willing to pay for non-value-adding activities. Unfortunately, typically 95% of process steps are non-value-adding, and only 5% add value to the product or service.

VSM also provides a common platform to apply various Lean principles and tools and allows an organization to create an integrated plan to follow for implementation. Furthermore, because it captures the delay caused by material and information flow on the same page, it is possible to tackle the two together, to create an optimized process such that the lack of one is not hampering the flow of the other.

The Benefits of VSM

VSM, used as a tool, provides several benefits to a process. The first main benefit is enabling people to see the entire process rather than just a single step. The flow of the entire process becomes apparent. This also makes the sources of waste in the value stream evident. With a clear process flow and identified sources of waste, the decisions needed to improve the flow are also apparent.

Another major benefit to the VSM technique is that it utilizes a format that provides a common language for the manufacturing process, which ties together Lean concepts and techniques. VSM is also the only tool that currently provides a link between information flow and material flow. VSM is conducive in any environment that contains a process to meet a desired objective.

As mentioned at the beginning of this chapter, VSM was taken from the TPS method called "Material and Information Flow Mapping." TPS used this method more as a means of communication by individuals learning through hands-on experience. It is used to illustrate the current and future states of a process to implement Lean systems. The focus at Toyota is to establish a flow, eliminate muda (the Japanese word for waste), and add value. Toyota teaches three flows in manufacturing:

1. The flow of material.
2. The flow of information.
3. The flow of people and processes.

VSM is based on the first two of these three: the flow of material and information.

Creating Current-State and Future-State Maps

The initial step is to map the current process. In evaluating the current state of a process, you can typically identify several improvements, including cellular manufacturing, one-piece flow, Single-Minute Exchange of Dies (SMED, covered in more detail in Chapter 13), and kaizen wherever possible. Next, you should map the improved process to represent the desired future state. The purpose of the current-state map is to make a clear representation of the production situation by drawing the material and information flows.

The goal of Lean is to get one process to make only what the next process needs when it needs it by linking the processes from raw material to the final customer. Value can be defined by the customer as a product that is delivered at the right time, with the defined specifications, and at the right price.

The purpose of VSM is to identify sources of waste and eliminate the waste by implementing a future-state value stream. The first pass of implementing a future-state value stream should ignore the inherent waste from product design,

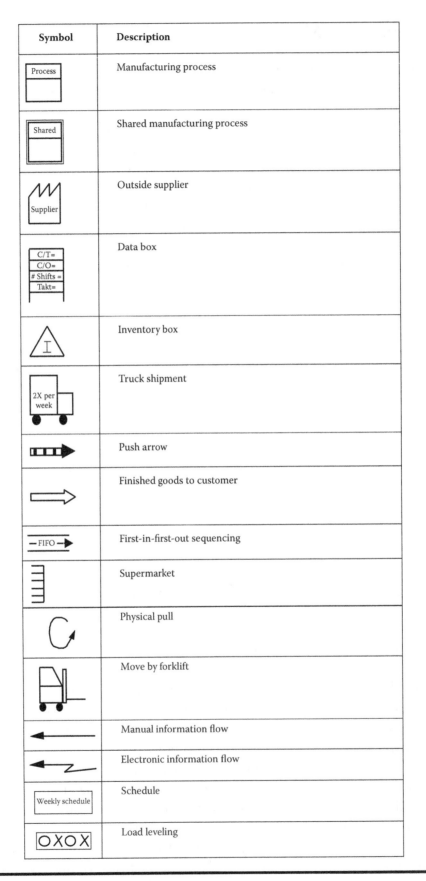

Symbol	Description
Process	Manufacturing process
Shared	Shared manufacturing process
Supplier	Outside supplier
C/T= C/O= # Shifts = Takt=	Data box
△ I	Inventory box
2X per week	Truck shipment
▦▶	Push arrow
⇒	Finished goods to customer
⟶ FIFO ⟶	First-in-first-out sequencing
▤	Supermarket
↻	Physical pull
	Move by forklift
←	Manual information flow
⟵⚡	Electronic information flow
Weekly schedule	Schedule
O X O X	Load leveling

Figure 6.2   Value stream mapping icons.

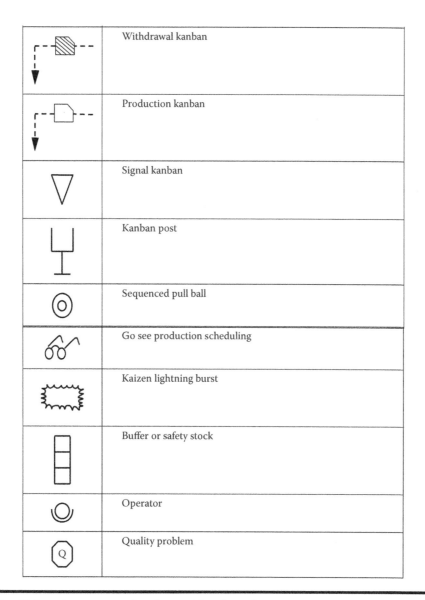

	Withdrawal kanban
	Production kanban
	Signal kanban
	Kanban post
	Sequenced pull ball
	Go see production scheduling
	Kaizen lightning burst
	Buffer or safety stock
	Operator
	Quality problem

Figure 6.2   Value stream mapping icons (Continued).

current processing machinery, and the location of some activities because these changes may require a great deal of work and will not be changed immediately. You should address these features in later iterations of your maps.

The common VSM symbols are given in Figure 6.2.

Conclusion

Now that we have reviewed the basics of VSM, let us revisit Carjo Manufacturing in the next chapter and see how Carjo has created current-state value stream maps for its manufacturing operations.

Conclusion

Chapter 7

Case Study: Creating Current-State Maps of Three Carjo Manufacturing Facilities

Now that we have revisited value stream mapping, let us take a look at how Carjo Manufacturing develops its current-state maps for its three plants. Later, in Chapter 8, the wastes and improvement opportunities will be identified using current-state value stream mapping for process improvement.

Calculating Takt Time to Understand Customers' Needs

Before starting the mapping process, the process improvement team first needs to understand its customers' requirements. It is vital for the process improvement team to understand the customers' requirements; however, every person in the organization should also understand the customers' requirements. *Takt time* is the frequency with which the customer wants a product. In other words, it is how frequently a sold unit must be produced. You can calculate this time by dividing your available production time in a shift by your customer demand for products that are made during a given shift. Takt time is usually expressed in seconds. However, depending on the product and the industry, takt time can be expressed in hours, days, or weeks.

For the gear-housing assembly at Carjo, the daily customer demand is 775 units. All three plants run a two-shift operation. Therefore, takt time can be calculated as shown in Figure 7.1.

Takt time is 67 s per gear-housing assembly. This information is critical as we gather our information for the current-state maps. All process operations must be less than takt time. In addition, as we balance the process operations later [in Part IV, specifically, in Chapter 11 on Six Sigma, Chapter 13 on Single-Minute

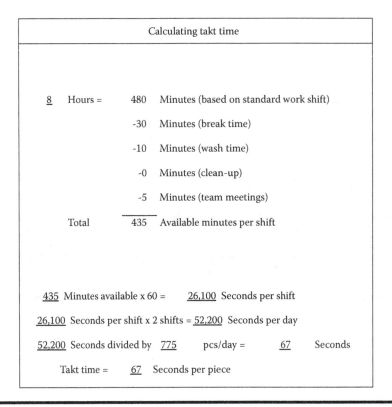

Figure 7.1   Takt time calculation.

Exchange of Dies (SMED), and Chapter 14 on Standard Work], we will need to load operators with this in mind.

Sample Current-State Maps for Carjo Plants

Figures 7.2, 7.3, and 7.4 show the current-state maps for Plants 1, 2, and 3, respectively. These provide the baseline for process improvements.

Conclusion

Now that Carjo has created its current-state maps, it is important to identify the improvement opportunities. The next chapter takes you through identifying the improvement opportunities at Carjo.

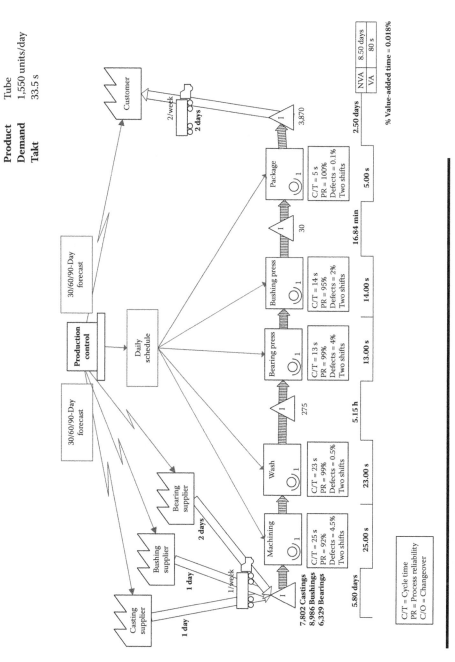

Figure 7.2   Plant 1 (tube) current-state map.

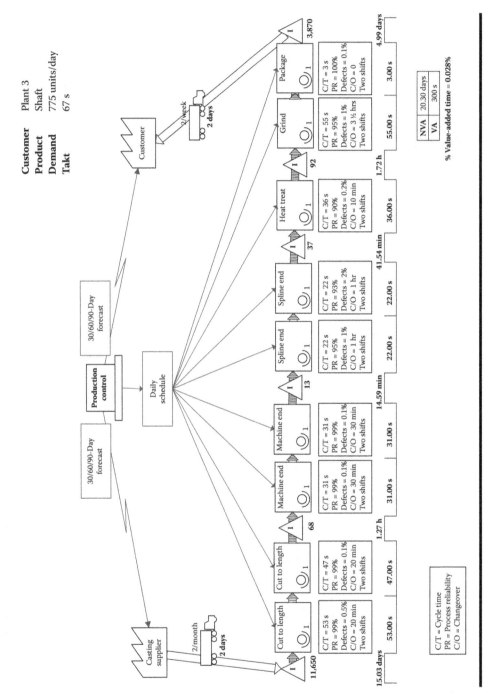

Figure 7.3   Plant 2 (shaft) current-state map.

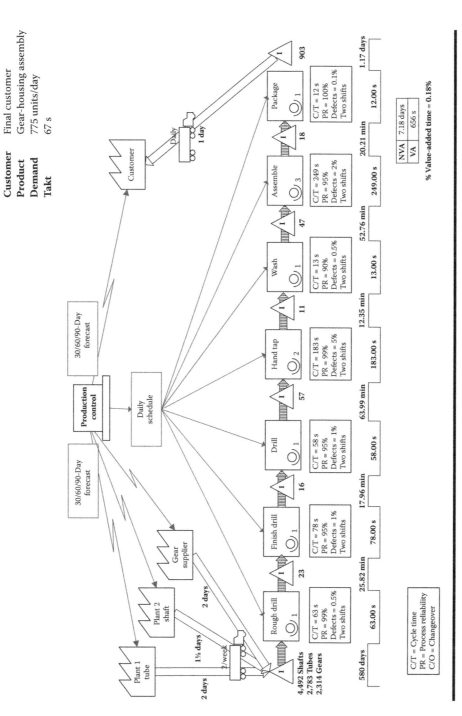

Figure 7.4 Plant 3 (gear-housing assembly) current-state map.

Chapter 8

Case Study: Identifying Improvement Opportunities at Carjo Manufacturing Co.

Now that Carjo's leaders have a picture of their company's current state and understand their customer demand, they can begin to identify areas for improvement. The current-state maps and takt time are essential to pinpointing bottlenecks that hinder flow and wastes.

There are several reoccurring issues with the processes in Plants 1, 2, and 3 that lead to excessive waste. Waste is any process or operation that adds cost or time and does not add value. The seven original wastes are described in Figure 8.1. An eighth waste of unused creativity should also be considered: we need to focus on our people and tap into their intellectual creativity.

Analyzing Waste in Carjo's Plant 1

Figure 8.2 shows the waste in the tube-operations processes in Carjo's Plant 1 using kaizen bursts (in the shaded areas). As you can see:

- The cycle times for all of the processes are considerably less than takt time.
- The value-added time is 80 s.
- However, the total lead time is 8.5 days, which calculates to a 0.018% value-add time for Plant 1.

The Toyota Production System (TPS) is considered best-in-class for implementing Lean techniques. TPS has several processes that have a percent value-added time as high as 35%; this is world-class performance. Companies typically range in the 3–5% range for percent value-added time when first beginning the

Wastes	Description
Overproduction	Producing more than what the immediate internal or external customer needs. Overproduction requires additional space, material handling, and storage that otherwise would not be needed.
Inventory	Inventory not immediately needed by the customer. This is typically caused by push scheduling.
Waiting	Time spent waiting for materials. This is typically caused by unbalanced production lines.
Transportation	Transportation does not add value because it does not contribute to transforming the final product. Point-of-use techniques can help minimize transportation waste.
Processing	Overprocessing waste can be caused by poor tool or product design.
Motion	Wasted motion includes double handling, reaching for parts, and stacking parts, to name a few examples. Point-of-use techniques can help eliminate wasted motion as well.
Defects	Poor quality of products requires production of additional products (which causes overproduction) to replace the defective parts, and it creates an inventory of unusable products. Poke-yoke techniques can help prevent defects from moving down the production line.

Figure 8.1   Seven original forms of waste.

Lean journey. In comparison, Plant 1 is operating at an extremely poor level of percent value-added time, which indicates there is considerable opportunity for improvement.

Analyzing Waste in Carjo's Plant 2

Figure 8.3 shows the current-state map for the shaft manufacturing process in Carjo's Plant 2; again, the waste in the processes are shown using kaizen bursts.

- All of the processes are operating well under takt time.
- The value-added time is 300 s, whereas the total lead time is 20.3 days.
- Therefore, the value-added time for Plant 2 is 0.028%.

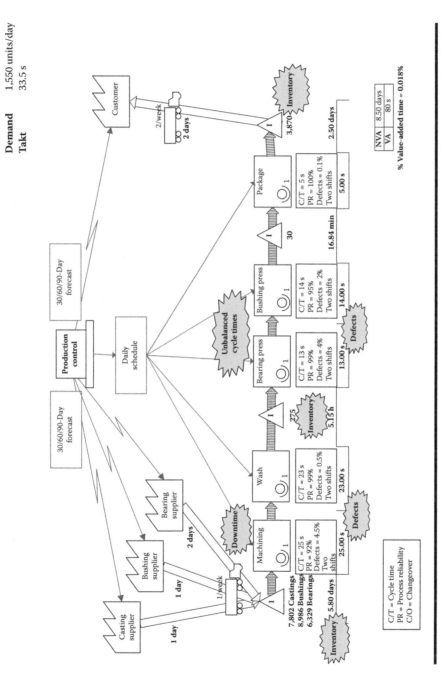

Figure 8.2 Plant 1 (tube) value stream mapping with wastes.

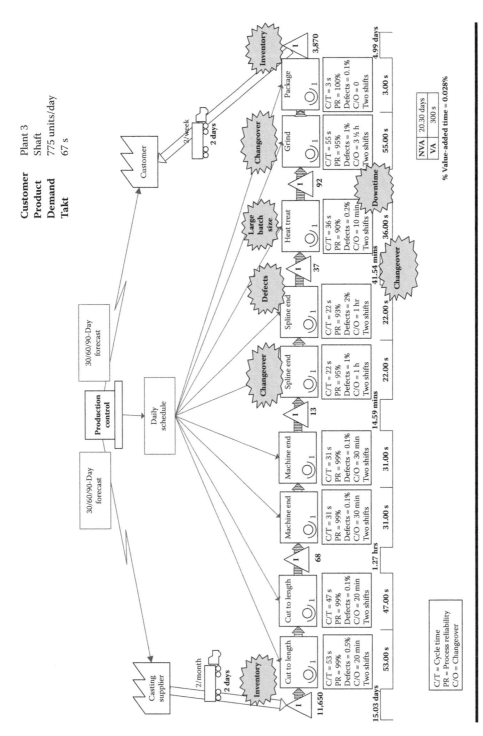

Figure 8.3 Plant 2 (shaft) value stream mapping with wastes.

The performance in Plant 2 is again well below the industry average.

Analyzing Waste in Carjo's Plant 3

Finally, Figure 8.4 shows the current-state map for Carjo's Plant 3, which machines the gear housing and performs the final assembly; again, it shows the waste in the processes using kaizen bursts. Here is what Figure 8.4 reveals about Plant 3:

- The value-added time is 656 s.
- The total lead time, however, is 7.2 days.
- That calculates to 0.18% value-added time.

Plant 3 has the highest percent value-added time of the three facilities, but it is still an extremely low percent.

In several cases, operators are assigned to a machine that is operating considerably below takt time, which is causing inventory to build up. This leads to the waste of overproduction and waste from idle/waiting time.

Given the takt time of 67 s, there are three operations in the gear-housing assembly process that cannot be completed within this time. These are bottleneck operations that cause other operations to wait. In addition, the long cycle times are leading to overtime to produce enough product for shipments, which increases production costs.

What Carjo Should Do to Eliminate Waste in All Three Plants

In addition to the problems at Plant 3, there is considerable inventory at each facility. The inventory exists mainly at the beginning and end of the process, because of infrequent deliveries. But there is also inventory built up as work-in-process throughout each facility caused by the lack of balance between operations.

For example, in Plant 2, the long changeovers for grind and spline operations are a significant disruption and waste in the process. The grind operation requires a 3.5-h changeover. Each spline machine requires a 1-h changeover.

So where should Carjo begin its Lean implementation? The company's previous method was to have each facility decide on the process improvement activities themselves. The kaizen event selection was up to the operations manager at each facility. No formal method existed to select the projects. The current-state value stream maps were used to drive the improvement project identification. But there was no clear method for prioritization of the projects. Previous projects resulted in gains, but in the overall process at the system level these gains did not have a big impact. To determine where Carjo should begin its

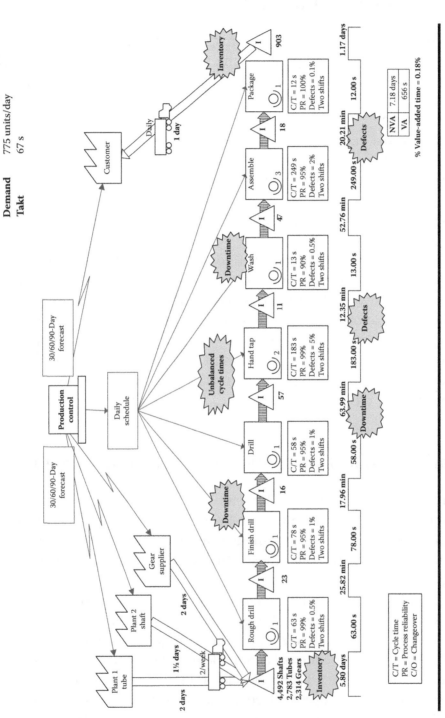

Figure 8.4   Plant 3 (gear-housing assembly) value stream mapping with wastes.

implementation, you first need to have a basic understanding of the process improvement tools, which are covered in Part IV: *Improve.* Then, Chapter 16 revisits Carjo to see how to cascade the strategic goals down into prioritized process improvement projects and daily activities.

Conclusion

Carjo Manufacturing has numerous areas for improvements, but how can it decide what process improvement tool to use? The next part of the book takes you through the most common process improvement techniques to enable you to select the most effective tool for each situation.

IMPROVE

At this point, you understand how to identify waste. Now you need to select the most appropriate tool for process improvement. Part IV takes you through the common process improvement techniques so that you can select the right tool to achieve effective improvements.

Chapter 9

Systems Thinking and Theory of Constraints

Now that Carjo's leaders have developed their strategic goals using Hoshin Kanri and understand their current situation using value stream mapping, you can begin the process improvements. At this stage, however, it is critical for you to consider the organization as a system of integrated processes. Some improvement efforts will only impact small portions of the business whereas other efforts can impact the entire business. The goal of this chapter is to get you thinking about the entire business as interconnected processes so that you can select improvement activities with the greatest impact on your organization.

Theory of Constraints (TOC) is a methodology that focuses on profit. The basis of TOC is that every organization has at least one constraint that limits it from getting more of whatever is the goal—typically, profit. TOC defines a set of tools that can be used to manage constraints.

Most organizations can be defined as a linked set of processes that take inputs and transform them into saleable outputs. TOC models this chain of linked processes. This system is based on the theory that a chain is only as strong as its weakest link.

Five Steps to Strengthening Your Weakest Link

Eliyahu Goldratt has defined a five-step process that a change agent can use to strengthen the weakest link.

- Step 1: Identify the system constraint. A constraint is anything limiting a system from achieving higher performance. This link can be either a physical or a policy constraint.

■ Step 2: Determine how to exploit this constraint. The change agent should obtain as much capability from the constraining link as possible. As with any other type of continuous improvement, these changes should be inexpensive measures.

■ Step 3: Subordinate the nonconstraint components. This will allow the constraint to operate at the level of maximum effectiveness. The overall system should then be reviewed to determine if the constraint has moved to another component. If the constraint is eliminated, the change agent skips Step 4 and continues at Step 5.

■ Step 4: Elevate the constraint. In this step, action must be taken to eliminate the constraint. These actions may include major changes to the existing system. This step is only necessary if Steps 2 and 3 were not successful.

■ Step 5: Return to Step 1. TOC is a continuous improvement process.

Three Ways to Measure Changes

TOC also defines three essential measurements to drive changes in the process:

1. *Throughput* is defined as the rate money is generated through sales of a product or service. This represents all of the money coming into an organization.

2. *Inventory* is the money invested in a product or service that an organization intends to sell. According to Goldratt, inventory includes all of the following:

   ■ Facilities
   ■ Equipment
   ■ Obsolete items
   ■ Raw material
   ■ Work-in-process
   ■ Finished goods

3. *Operating expense* is the third measurement; it is defined as the money used to turn inventory into throughput. Operating expenses include items such as:

   ■ Direct labor
   ■ Utilities
   ■ Consumable supplies
   ■ Depreciation of assets

The three measurements are interdependent on each other. Because one measurement will change one or both of the other measurements, TOC has defined the following formula for the change agent:

Maximize throughput while minimizing inventory and operating expense.

All improvement efforts should be prioritized by how each affects the three measures. Therefore, throughput is especially critical because the only limit on how much it can be increased is the market size.

Conclusion

Systems thinking is critical to making process improvements that impact the overall organization rather than a subset. You need to think about the big picture and focus on improvements that are tied to the strategic objectives of the organization. The next chapter provides an overview of Lean tools and their progression over time.

Chapter 10

Tool Selection

Using systems thinking from the previous chapter, you now need to select the most appropriate tool to alleviate your systems constraint. This chapter takes you through the history and progression of Lean, which often mirrors the process improvement milestones you will undertake in your own efforts. Chapters 11–15 take you through the common process improvement tools in detail.

Lean emphasizes the elimination of waste and creation of flow within an enterprise. The primary focus of Lean is on the customer, to address value-added and non-value-added tasks. *Value-added tasks* are the only operations for which the customer is ready to pay. The idea in creating flow in Lean is to deliver products and services just in time, in the right amounts, and at the right quality levels at the right place. This necessitates that products and services are produced and delivered only when a pull is exerted by the customer through a signal in the form of a purchase. A well-designed Lean system allows for an immediate and effective response to fluctuating customer demands and requirements.

An Overview of Lean Tools

Lean tools that are most commonly used to eliminate waste and achieve flow are as follows:

- Value stream mapping (which we have already discussed in detail in Chapter 6)
- Six Sigma (6σ; discussed in detail in Chapter 11)
- 5S housekeeping (discussed in detail in Chapter 12)
- Visual management (discussed with 5S in Chapter 12)
- Single-Minute Exchange of Dies (SMED; discussed in detail in Chapter 13)
- Standard Work (discussed in detail in Chapter 14)
- Mistake-proofing (discussed in detail in Chapter 15)

By linking our improvements to our long-term strategy, we can select the appropriate improvement projects and, in turn, select the appropriate tool that will impact the system as a whole. The purpose of this chapter is to illustrate the progression of Lean over time, which will be similar to how Lean will evolve within your processes. There are different rates of change. Kaizen, as you will see, is small incremental changes. However, kaikaku is a rapid change. On the basis of your current processes and the speed of your improvement events, you will likely see similar progression, as described in the following sections.

Lean History and Progression

Manufacturing has continued to evolve throughout time from traditional manufacturing to Heijunka, with several steps in between. Each period had various methods of controlling production instruction, production control, material, man, machine, and lead time.

Prior to the 1900s, products were handcrafted. Typically, they were made one at a time from start to completion. Therefore, delivery was typically slow. Quality was high because the focus was on that one product. However, there was a wide variation in product and process because a process step from one product to the next product might be several weeks apart.

Batch production was used in Henry Ford's assembly line. There was one operator at each machine or station and therefore a single-task orientation. The machines were high volume with a focus on high machine utilization. Subsequently, there were large production lots of sequential steps; however, there was little variation in the process or product. Lead times in batch production are long because each product in the batch must wait for the other products in that batch to be completed before it can continue to the next process. The next sections describe each manufacturing environment up to current practices.

The Traditional Manufacturing Environment

Traditional manufacturing is characterized by verbal production instructions. The production build is forecasted and changed often. Product is made at many locations and is scheduled. Inventory and material is decentralized and difficult to control. Material is bought and stored in large nondefined quantities. Processes are individualized for specialized operators. Production is aimed toward maximum capacity and machine utilization. The equipment is dedicated with long setup times. The lead time to manufacture a product is highly variable and difficult to predict.

The Continuous Flow Manufacturing Environment

Continuous flow manufacturing is similar to traditional manufacturing in that the production orders are verbal, forecasted, and change often. Product in continuous flow manufacturing is manufactured at limited locations. Production is scheduled and decentralized. Focus is now on controlling the processes. Material has changed from large undefined quantities to limited work in process. The operator responsibilities have also changed from specialized tasks to related tasks within processes. Equipment is now arranged according to the process flow rather than grouped by function. Lead time has become more predictable and is reduced.

The Standardized Work Environment

In a standardized work environment, production instruction is similar to continuous flow with verbal instructions, forecasting, and limited production locations. Product is manufactured to customer order with a defined work in process. The process is well defined to the work sequence for the operator. Machines are synchronized to approximately the same process speed, and the lead time for a product is predictable.

The Pull System of Manufacturing

The pull system is based on a concrete order for the customer requirements. Production control is visible and disciplined, typically with the use of kanbans. Material is also controlled with the use of kanbans to replenish the system and determine proper inventory levels. With systems to control inventory, the operator's indirect work becomes manageable. Lead time now is based on the customer rather than the process. The customer lead time is predictable.

Manufacturing Environments That Use Small Lots

The method of using small lots is similar to the pull system; however, customer requirements change frequently. Material requirements are reduced because of the reduced size of the lots. Kaizen is used to improve the processes and reduce the lead time. The equipment is frequently changed over for different products. SMED is used to reduce changeovers. With the use of kaizen and SMED, the lead time for an order is reduced.

The Heijunka Method of Leveling Production Work

Heijunka is a method focused on leveling production. Production instruction and control is similar to that of the small-lot method. Material usage is uniform with a leveled production. The operator is also leveled with respect to his or her workload, and machine utilization is uniform. Again, the lead time for production is reduced and predictable.

Just-in-Time Manufacturing

The Just-in-Time (JIT) production system is based on the philosophy to provide the right product or the right service in the right quantity or amount at exactly the right time on the basis of customer requirements. JIT drives to eliminate waste. The seven forms of waste are overproduction, transportation, inventory, overprocessing, waiting, motion, and defective parts. An eighth waste of unused creativity can also be considered. (Figure 8.1 in Chapter 8 provides a more detailed description of each of the seven original wastes.) By implementing JIT concepts to eliminate waste, JIT drives improvements in quality, delivery, and cost.

The key to the JIT system is the customer. The customer in the past could only select high quality, good service, or low price. At best, the customer might be able to get two out of the three. Now in order to satisfy customer demands, you must provide for all three. Therefore, the profit equation has changed drastically over time. The old calculation was to add the cost to the profit desired to determine your selling price: *cost + profit = selling price*. In contrast, the new method is to calculate profit by subtracting cost from the selling price: *profit = selling price − cost*.

One-Piece Flow and Cellular Manufacturing

Taiichi Ohno at Toyota developed Lean production, in which operators are multi-skilled and run several machines or steps. Lean production may incorporate cellular design; however, the flow production is based on the customer order. Other characteristics include one-piece flow and flexible setups. One-piece flow in a cellular layout has many benefits:

- It significantly reduces transportation, inventory, and waiting time.
- At the same time, one-piece flow will significantly improve quality, delivery, and cost.
- It lowers lead times.
- It improves product distribution.
- It reduces scrap and rework.
- It makes scheduling easier.
- It uses floor space better.
- It reduces material handling.
- It uses labor better and increases productivity.
- It exposes problems.

Cellular manufacturing is a group of machines or processes connected by process sequence in a pattern that supports the efficient flow of production. There are several elements that are characteristic of a good cell design:

- The process will determine the layout.
- Quality is built into the process at each step rather than at a final inspection.
- One operator could run the cell.
- Machines are in close proximity to each other.
- The cell is a U-shaped design that flows counterclockwise.
- The operators are multiskilled.

The cell design should focus on the operator. This is based on the philosophy that people *appreciate* over time, whereas machines *depreciate*. In other words, people are more valuable than machines.

The cell should also be designed around the process. Similarities between products can be determined by using Part/Quantity Analysis to show model/volume relationships and Process Route Analysis to show process relationships.

Kaizen

Kaizen is the philosophy in which one seeks to continuously improve. It is a state of mind to never accept status quo. *Kai* means change, and *zen* means for the good. Kaizen is a constant process. What is done today to improve should be done tomorrow as second nature. It drives us to improve. Kaizen should also be done with the imagination before turning to costly improvements.

The improvement cycle begins with exposing and quantifying problems. The next step is determining root cause. After uncovering the root cause, solutions should be implemented. After implementation, the focus is on standardization and adherence. Because this is a cycle, the process continues for further improvements.

Conclusion

Now let us take a closer look at the Lean Six Sigma toolbox. This will enable us to select the appropriate technique in the next step of Hoshin planning as we move into daily management activities. Chapters 11–15 each describe one tool in detail; let us begin with Six Sigma, the subject of Chapter 11.

Chapter 11

Using Six Sigma to Improve Quality

The first Lean tool we will discuss is to help reduce variation and improve quality in our processes is Six Sigma. Six Sigma is covered first so that you can focus on improving external quality for your customer. If you have a high defect rate for product reaching your customer, then your customer will be extremely dissatisfied. This will likely hurt your business more than slow production cycles or long changeovers, so it is important to focus first on sending high-quality products out of your facility.

Six Sigma is a powerful tool for improving quality, productivity, profitability, and market competitiveness. It is focused on reducing variation using a problem-solving approach and statistical tools. Six Sigma is a customer-focused continuous improvement strategy and discipline that minimizes defects and variation toward an achievement of 3.4 defects per million opportunities in product design, production, and administrative processes. It is focused on customer satisfaction and cost reduction by reducing variation in processes. Six Sigma is also a methodology using a metric based on standard deviation.

Goals and Benefits of Six Sigma

The goals of Six Sigma include:

- Developing a world-class culture
- Developing leaders
- Supporting long-range objectives

There are numerous benefits of Six Sigma, including:

- Stronger knowledge of products and processes
- Reduction in defects
- Increased customer satisfaction level that generates business growth and improves profitability
- Increased communication and teamwork
- A common set of tools

Five Phases of Six Sigma Strategy

There are five key phases in the Six Sigma strategy, abbreviated as DMAIC.

1. Define
2. Measure
3. Analyze
4. Improve
5. Control

The following sections describe each phase in more detail.

Phase 1: Define the Process

The purpose of the first phase of the DMAIC process is to identify and refine a process in order to meet or exceed the customer's expectations. The Define phase includes developing all of the following:

- The team charter
- Critical-to-quality characteristics
- A problem statement
- A communication plan
- The project scope
- Goal statements

In Phase 1, you should ask the questions listed in Figure 11.1.

Phase 2: Measure the Process

The purpose of the Measure phase is to develop, execute, and verify a data collection plan. The tools typically used in the Measure phase are as follows:

- Who are the team members?
- Have the team members been properly trained?
- Is the team adequately staffed with the desired cross-functionality?
- Have you identified the customer(s)?
- Has the team collected the voice of the customer?
- Have the customer needs been translated into measurable requirements?
- Has the team developed and communicated the project charter?
- What specifically is the problem?
- What is the project scope?

Figure 11.1   Questions to ask in Phase 1 of Six Sigma: Defining the process.

- Process flow diagrams
- Process failure mode and effects analysis (FMEA)
- Measurement system analysis (MSA)

In Phase 2, you should ask the questions listed in Figure 11.2.

Phase 3: Analyze the Process

The purpose of the Analyze phase is to develop and test hypotheses about the causes of process defects. Therefore, hypothesis testing is a common tool in this phase. In the Analyze stage, there are three main questions, listed in Figure 11.3.

- What processes are involved?
- Who is the process owner?
- Which processes are the highest priority to improve?
- What data support the decision (i.e., what metric)?
- How is the process performed?
- How is the process performance measured?
- Is your measurement system accurate and precise?
- What are the customer-driven specifications for the performance measures?
- What are the improvement goals?
- What are the sources of variation in the process?
- What sources of variability are controlled and how?

Figure 11.2   Questions to ask in Phase 2 of Six Sigma: Measure the process.

■ What are the key variables that affect the average and variation of the performance measures?

■ What are the relationships between the key variables and the process output?

■ Is there interaction between any of the key variables?

Figure 11.3   Questions to ask in Phase 3 of Six Sigma: Analyze the process.

■ What are the key variables settings that optimize the performance measures?

■ At the optimal setting for the key variables, what variability is in the performance measure?

Figure 11.4   Questions to ask in Phase 4 of Six Sigma: Improve the process.

■ How much improvement has the process shown?

■ How much time and/or money was saved?

■ Long-term metric?

Figure 11.5   Questions to ask in Phase 5 of Six Sigma: Control and monitor the process.

Phase 4: Improve the Process

The Improve phase focuses on formulating and implementing process improvement ideas. Tools in this stage include Multiple Regression and Design of Experiments (DOE). A good source for more information on these tools is *Six Sigma Fundamentals* by D.H. Stamatis. In the Improve phase, you should ask the questions listed in Figure 11.4.

Phase 5: Control the Process

The final phase of the DMAIC process includes controlling and monitoring the process. In this stage, you need to address the questions listed in Figure 11.5.

Conclusion

Six Sigma is a powerful tool for reducing variation and improving quality in processes. Now let us take a look at a tool for improving workplace organization and cleanliness. Chapter 12 discusses 5S in detail to help you create an organized work environment.

Chapter 12

Using 5S to Create a Clean and Manageable Work Environment

Whereas Six Sigma (described in Chapter 11) is a Lean tool that improves product quality by reducing process variation, 5S is another Lean tool that improves quality and productivity in a different way. 5S builds a foundation for a company to delivery high-quality products and services in the right quantity at the right time to satisfy and exceed customer requirements. It is key to creating a work environment that focuses on quality. 5S also creates a safer and more pleasant place to work. It makes it possible to increase productivity and improve quality. The term 5S is derived from five Japanese terms for practices that lead to a clean and manageable work environment.

Overview of 5S

There are five steps in 5S, defined briefly here and discussed in more detail in the rest of this chapter.

- Step 1: Seiri, or *simplify*. In this step, the team must determine what is necessary and unnecessary. The unnecessary must then be disposed of.
- Step 2: Seiton, or *straighten*. The team organizes the necessary items so they can be found easily, used, and then easily returned to the proper location.
- Step 3: Seiso, or *scrub*. This not only includes cleaning the equipment, but also cleaning the floor and furniture in all areas of the workplace.
- Step 4: Seiketsu, or *stabilize*. In this step, the first three steps are maintained and improved.
- Step 5: Shitsuke, or *sustain*. The discipline or habit of properly maintaining the correct 5S procedures is maintained.

In order for a 5S program to be successful, it must be fully supported by management. They must give guidance, coordination, support, and proper communication. Management must provide a suitable environment for employees to utilize their skills properly. An important element is management's support to encourage the team to focus on value-added activities to expose problems and respond accordingly.

Benefits of 5S

There are numerous benefits from implementing 5S principles.

■ Quality is improved by clearly identifying all necessary objects in an area.
■ 5S reduces waste by eliminating unnecessary steps to search for necessary materials, tools, or equipment.
■ Safety is also improved by providing a place for everything and having everything in its place.
■ Maintenance is easier by having all of the needed tools in a specific location.
■ 5S improves profitability by reducing waste, improving quality, and eliminating all unnecessary equipment.

5S also provides benefits that directly relate to the employee.

■ Employees take more pride in their workplace when the area is clean and organized.
■ Respect is gained from associates and customers after they visit the area.
■ Employees also keep a more positive mental attitude working in this environment.
■ 5S can build trust, remove frustration, and improve morale.

The 5S Steps

Step 1: Seiri—*Simplify the Workplace*

The first step of 5S is to simplify (*Seiri*). In the first step, all items in the work area should be sorted. The team clearly distinguishes between items that are necessary and items that are unnecessary. The unnecessary items are then disposed of. *Seiri* means separating the necessary equipment such as tools, parts, and instructions from the unneeded materials and disposing or removing the latter.

An example of how simplifying improves quality can be seen with two different but similar parts that are used to make two different products. Occasionally, the wrong part was used in assembly. The parts were color-coded along with their container and work order to reduce confusion. Simple color-coding can be an easy way to eliminate unnecessary confusion in daily activities.

This step is essential because it removes waste, which improves quality. It also makes the workplace safer by eliminating clutter. Floor space is gained by disposing of all of the unnecessary items. This makes the necessary items that remain much easier to visualize. Simplifying the area reduces crowding and clutter, which improves the workflow. The wasted time searching for tools is eliminated. Decreasing the unneeded inventory and equipment also reduces costs. Excess stock can hide other problems such as quality defects. Unneeded items make improving process flow difficult.

Separating Necessary from Unnecessary Items

The team must thoroughly examine the area. All unnecessary items should be thrown away. Items thought to be necessary should be considered for their use, why it is used, and how often it is used. In a production area, your team should evaluate such items as components, documentation, supplies, tooling, gauges, parts, and machines. When using 5S in an office area, your team should evaluate such items as records, forms, books, supplies, equipment, and parts.

Using Red Tags for Identification

An important part of the simplify step is red-tagging any items that are unnecessary or in question. The red tag should be a large piece of red paper, typically 8.5" × 11". The lower portion of the tag should be perforated to keep as a record of the location of the tag. An example is provided in Figure 12.1.

To organize the red-tag materials, create a red-tag board for tracking purposes. Then move all red-tagged items into a temporary red-tag area, and set up a time to dispose of the items. This will allow items to be removed by other teams for their use instead of them buying new material.

After the predetermined time period is up, dispose promptly of all remaining items. Continue this process on a regular basis to ensure that more unnecessary items do not begin to clutter the area again.

The steps in a red tag effort include:

1. Identify red-tag target/area.
2. Establish a criterion for evaluation.
3. Make the red tags.
4. Attach tags and separate lower portion for traceability.
5. Evaluate the tagged items.

Step 2: Seiton—Straighten Up the Workplace

In the second step, the entire workplace is organized/straightened. *Seiton* means arranging and identifying parts, materials, and tools for ease of use. Items should be placed in the best location for point of use and visually organized. This

Area:	Red tag	Tag #:

Category (circle):

Supplies	WIP	Tools
Office materials	Raw material	Equipment
Furniture	Finished goods	Other
Books/magazines/files		_____

Date tagged:	Tagged by:

Item:	Quantity:

Reason:

Disposition (circle)

Diacard	Sell
Store in dept.	Transfer
Long-term storage	Other:_____

Action taken:	Date:

- Perforated line -

Area:	Red tag locator	Tag #:

Figure 12.1 Red tag.

improves quality by visually identifying all products and creating a specific location for that item, which makes it easier for employees to find, use, and easily return all items to their proper location.

First, the team should decide where items should be placed for easier organization, and the locations and systems should be readily understood. When selecting a storage method, the team should pay careful attention to detail;

their goal should be to minimize inventory and space and improve visual management of items. Again, the locations of items are visually indicated so items can be easily returned. This also promotes readily identifying missing items.

Here is a list of items that should be identified using visual controls.

■ Shadow boxes for tools
■ Tooling/fixtures
■ Movable objects
■ Documentation control
■ Storage area for common tools
■ Raw material
■ Finished material
■ Discrepant material
■ Packaging material

Step 3: Seiso—*Scrub the Workplace*

In the third step, all of the areas of the workplace are thoroughly cleaned, including floors, equipment, and furniture. *Seiso* means performing a cleanup campaign, which is important because a clean work environment promotes quality work.

While cleaning, the team should consider how the area got dirty to help maintain a clean environment. The team should inspect for safety hazards or leaks, and before continuing with the cleaning, the team should repair any items that need to be fixed. Cleaning equipment should also have an identified location to ensure that it can be found readily. These activities should be integrated into daily maintenance.

The areas in the workplace that should be cleaned include:

■ Floors
■ Ceilings
■ Walls
■ Computer equipment
■ Furniture
■ Cabinets
■ Desks
■ Production equipment
■ Unnecessary computer files

Step 4: Seiketsu—*Stabilize the Workplace Standards*

Step 4 consists of maintaining and improving the first three 5S standards by implementing needed changes. *Seiketsu* means performing *Seiri, Seiton,* and *Seiso* at specified frequency intervals to maintain and improve the well-organized and clean work environment. This is an essential step in any type of quality improvement effort.

Seiketsu can be performed by creating a daily checklist of cleaning and organizing activities. You can also perform 5S patrols and spot checks to monitor the progress of the 5S efforts. During this step, you should make changes to existing equipment to make cleaning quicker and easier, and you can use the Five-Why method to determine the root source. You can then modify equipment and systems to improve the cleanliness of the area. You should also eliminate flat storage areas and excess storage areas.

During Step 4, you should also schedule time and resources for 5S activities. You can create model areas to demonstrate the 5S philosophy, and you can use checksheets to provide feedback. You should arrange for management walk-throughs to be performed to maintain the 5S culture. Finally, you should require training on 5S to occur on a specified schedule.

Step 5: Shitsuke—*Sustain*

In the final step of 5S, you must instill the discipline or habit of properly maintaining the 5S procedures. 5S must become a habit in your team's daily life. *Shitsuke* means forming a habit of following the first four S's. The team must develop a habit of simplifying, straightening, scrubbing, and stabilizing the work environment. The entire team must be focused on the goal of 5S.

This is the most difficult "S" to obtain. It means maintaining self-discipline and practicing 5S until it is a way of life. Focusing on continuous improvement is essential. An important part of any 5S program is also sharing the lessons learned. By sustaining the effort, employee spirit is improved. Management involvement is critical for the success of the 5S effort to continue to grow and sustain.

Using Visual Boards to Document 5S Status

There are several forms that you can use to visually document the 5S status of your workplace.

1. The first form is the 5S wheel, which gives a visual overview of the level of 5S an area has obtained. As each level is obtained, that "S" is filled in, and levels not yet achieved are left unfilled. Figure 12.2 shows an example of a 5S wheel.
2. The second form is a series of questions that the team should address to obtain each level. Figure 12.3 is a checklist of the types of questions that can be asked for each "S." Post these forms in each area on the shop floor.

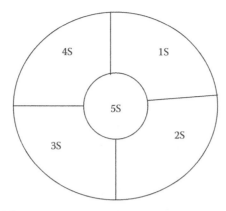

Figure 12.2   5S wheel.

Conclusion

5S is a powerful tool for improving workplace organization. You can use the forms provided in this chapter to implement 5S in your workplace. Chapter 13 takes you through the next improvement tool, Single-Minute Exchange of Dies (SMED), which improves changeover time.

5S Evaluation sheet	Year:	Area:																							
Manufacturing/shop floor		Jan		Feb		Mar		Apr		May		Jun		Jul		Aug		Sep		Oct		Nov		Dec	
		Y	N	Y	N	Y	N	Y	N	Y	N	Y	N	Y	N	Y	N	Y	N	Y	N	Y	N	Y	N
1S Simplify (Seiri) – Organize needed from unneeded, remove the unneeded																									
1	Are only necessary materials or parts present?																									
2	Are defective materials identified and segregated?																									
3	Are unused machines, equipment, or parts marked with a red tag?																									
4	Are red tagged items in the red tag area?																									
5	Is standard work present and current?																									
2S Straighten (Seiton) – Arrange neatly, identify for ease of use																									
6	Are areas and walkways clearly marked?																									
7	Are materials visually controlled with location indicators and at the point of use?																									
8	Are maximum and minimum allowable quantities indicated?																									
9	Is nonconforming material segregated, identified, and documented?																									
10	Are equipment and tooling clearly identified and arranged in order of use?																									
3S Scrub (Seiso) – Conduct a cleanup campaign to polish and shine the area																									
11	Are floors and areas clean and free of trash, liquids, and dirt?																									
12	Is equipment inspection combined with equipment maintenance?																									
13	Is the work area lighting sufficient, ventilated, and free of dust and odors?																									
14	Are cleaning operations assigned and visually managed?																									
15	Are cleaning supplies easily available and stored in a marked area?																									
4S Stabilize (Seiketsu) – Continue the habits frequently and often																									
16	Are standard procedures clear, documented, and actively used?																									
17	Are there records of defective material entries with corrective action?																									
18	Are equipment TPM procedures in place and current?																									
19	Are waste and recyclable material receptacles emptied regularly?																									
20	Are scrap and product nonconformance areas cleared at a set interval?																									
5S Sustain (Shitsuke) – Institute the 5S practices as a part of the culture																									
*	The 5th S is awarded when a minimum of 3 consecutive months of 1S, 2S, 3S and 4S have been sustained.																									

Figure 12.3 5S checklist.

Chapter 13

Using Single-Minute Exchange of Dies to Reduce Setup Time

In addition to the 5S process of creating a clean and manageable work environment, another Lean tool that improves the workplace is Single-Minute Exchange of Dies (SMED), a methodology that focuses on reducing setup time. The ultimate goal of SMED is zero setup, in which changeovers are instantaneous and do not affect continuous flow. The main benefits of SMED (which are described in more detail in this chapter) are as follows:

- Reduced inventory
- Improved flexibility
- Increased capacity
- Better customer service

In short, the purpose of SMED is to make setup a standard, routine operation.

Origin of SMED

Shigeo Shingo introduced SMED after 19 years of observing setup operations in factories. It was initially developed to improve die press and machine tool setups. Setup time is calculated as the time between the last good piece of product A being completed until the first good piece of product B is completed. This includes the time to change materials, tools, fixtures, dies, etc.

In developing his methodology, Shingo discovered two main points. First, setup operations can be divided into two categories:

1. Internal setups, which are performed while the machine is not operating. An example is changing the tooling in the machine.

2. External setups, which are performed while the machine is running. An example of this is setting tools for a different product.

The second point Shingo discovered is that converting internal setups to external setups can substantially reduce changeover times. This also increases the time the machines can operate. Shigeo Shingo believed any setup time could be reduced by 59/60ths.

Benefits of SMED

As mentioned, implementing SMED provides for many benefits:

1. There are fewer adjustments, which mean less of a chance for error.
2. Flexibility in scheduling is increased because setup is less than takt time.
3. The expense of excess inventory is avoided. This, in turn, increases capacity by reducing the amount of lost time for changeover.
4. There is less material waste lost in setup.
5. Product quality is improved by reducing variation between each setup.
6. By producing smaller runs, there is a decreased chance of a large-scale defect in inventory.
7. SMED reduces defects from setup errors and eliminating trial runs.
8. Customer service is improved because of the ability to changeover quickly and meet changing customer needs. SMED enables companies to produce more cost-effective products with less waste in smaller quantities.
9. Faster changeovers increase productivity by reducing the amount of downtime.
10. SMED results in a safer work environment. By simplifying setups, the risk of physical strain or injury is reduced.
11. Fewer inventories reduce the amount of clutter.

SMED Methodology

All setup operations consist of five main steps:

1. Before beginning SMED, the setup operation must be determined.
2. The next step includes preparation, after-process adjustments, and checking of materials and tools.
3. The third step is mounting and removing blades, tools, and parts.
4. Next, measurements, settings, and calibrations are performed.
5. The final step is trial run and adjustments.

Let us look at each step in more detail.

Step 1: Determining the Setup Operation

In observing the current setup, the first step is to select the area for improvement. In selecting an area, do not think of the machines as independent from each other. Instead, you should consider setup-time reduction as part of a complete flow of production. The manufacturing process should consist of the continuous flow from raw materials to finished product. This includes the four basic phases of manufacturing processes: processing, inspection, transport, and storage.

Step 2: Preparation, After-Process Adjustments, and Checking of Materials and Tools

The purpose of this step is to ensure that all necessary tools and materials are in place and functional. You should perform this step external to the machine operating. This is the primary source of streamlining the setup process.

Step 3: Mounting and Removing Blades, Tools, and Parts

This step consists of removing parts and tools after processing a part and then attaching the parts and tools for the next part. To perform this step, typically you must stop the machine; therefore, it is an internal setup.

Step 4: Measurements, Settings, and Calibrations

In this step, you should perform the necessary measurements and calibrations during production operations. These operations are typically internal, but they can be converted to external.

Step 5: Trial Runs and Adjustments

The final step in a changeover is to make adjustments after machining the first piece. If you have taken accurate measurements and calibrations during the previous step, that will make these adjustments easier. During traditional setup, this is considered to be an internal setup element because changeover is not complete until good parts are produced.

Analyzing Setup

There are three stages of SMED that simplify changeover.

1. Separating internal and external setup.
2. Converting internal setup to external setup.
3. Streamlining the setup operation.

The three stages are outlined in Figure 13.1.

The preliminary step to implementing SMED improvements is to analyze the current operation. First, observe the setup by using videotape for easier reference. Pay attention to each machine operator's hand, eye, and body movements. Then show the video to the setup person and others involved with the operation. Finally, using the video, record the time and motion on a setup analysis chart, as shown in Figure 13.2. Then chart each time category in Pareto format. An example of the format is provided in Figure 13.3. There are several time categories, including searching, fixture change, walk time, first-piece inspection, gaging, tool change, and programming.

The key to successfully implementing SMED is to properly distinguish between internal and external setups. You should perform elements such as preparation and transport while the machine is running; this can typically reduce internal setup time by 30–50%. Another method of reducing internal setup is to reexamine operations that are assumed to be internal. You may assume that some steps are internal but they can be performed externally. You can also convert internal setup to external setup by identifying the function of those steps during the setup process. Another means of further reducing setup time is to analyze the internal setup steps to shorten the time needed. The next sections of this chapter describe the three steps of analyzing setup in more detail.

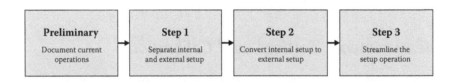

Figure 13.1   Stages of SMED implementation.

| Setup analysis chart | Machine from: 1000 / A | To B | Date |

Step #	Changeover element	Time Element	Internal	External	Waste	Improvement plan	Eliminate	Internal to external	Reduce
1	Remove fixture (fixture change)	12	12			Eliminate bolts —install quick release			X
2	Search for tools (search)	8	8		X	Mount tools at machine	X		
3	Clean fixture (fixture change)	10	10			Standardize procedure to externalize step		X	
4	Put on fixture (fixture change)	17	22			Eliminate bolts —install quick release			X
5	Load tools (tool change)	15	15			Standardize tooling on all models	X		
6	Machine first part (first piece)	3		3					
	Totals	65	62	3					

Figure 13.2   Setup analysis chart.

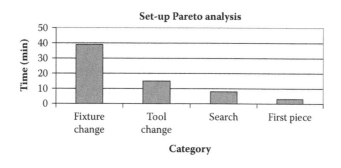

Figure 13.3   Setup Pareto analysis.

Step 1: Separate the Elements

You must separate internal and external elements in order to identify areas for improvement. The goal of SMED is to have all of the necessary tools at the station so the operator never leaves (and never needs to leave) to perform any of the external setups. You should also identify the separation of internal and external elements on your setup analysis chart under changeover categories, as shown in Figure 13.2. Internal activities are performed while the machine is not operating. External elements are performed while the machine is running.

Your machine operators can easily perform several tasks while the machines are running; for example, they can call the proper personnel, set tools, and get parts. However, many operators often do not begin these tasks until the machine stops. As mentioned earlier in this chapter, by performing these tasks external to the setup, you can typically reduce setup time in your organization by 30–50%.

Practical techniques to separate internal and external tasks include checklists, function check, and improving transport.

Using Checklists

Your checklists should include everything required to set up and run the next operation, including tools, specifications, people, operating conditions, and measurements. By checking off items on the list before the machine is stopped, operators can correct missed steps and errors prior to the internal setup. The checklist should also be specific for each machine or operation, because general checklists can be confusing and are therefore often ignored.

Using Function Checks

Function checks are also important before beginning the internal setup because they verify that the parts are in working order, and they allow time for repair before changeover. If your operators do not find broken equipment until setup, there may be a large delay during internal setup.

Improving Transport

Another means of reducing the amount of time the machine is down is to move the parts and tools to the machine during external setup. Machine operators should transport all necessary tools and equipment to the machine or operation before they shut down the machine for changeover. On the other hand, employees should not move the parts and tools from the previous operation to storage until they have installed the new parts and the machine is running.

Step 2: Convert Internal Setup to External Setup

In the previous step, you separated the internal tasks from the external tasks. To reduce the setup time, you need to convert internal setup to external setup. The first stage is to evaluate the true functions and purposes of each task in the current internal setup. The second stage is to convert the internal steps to the external setup. There are three methods to convert internal setup to external setup, described in the following paragraphs.

Technique 1: Prepare the Operation Conditions in Advance

This means having the parts, tools, and conditions ready before starting the internal setup. An example is preheating the die molds in advance, instead of heating the molds after the setup.

Technique 2: Standardize the Essential Functions

This means keeping characteristics the same from one operation to another. If tools or parts vary between operations, the operators must typically make adjustments while the machine is down, which tends to be very time consuming. In contrast, by standardizing the parts and tools, you can reduce internal setup time.

Functional standardization focuses on standardizing only the elements essential to the setup, such as securing the mold or fixture, centering, and dimensioning. This involves two steps:

1. First, evaluate each function in the setup process and determine which can be standardized.
2. Next, evaluate each function again to determine if a more efficient means is possible by using fewer parts. An example is standardizing the tooling for each model in order to eliminate changing the tools during setup.

Technique 3: Use Intermediary Jigs

These are standard dimension plates or frames that can be removed from the machinery. The purpose is to externalize as much of the setup as possible. The current fixture is attached to an intermediary jig on the machine. The next fixture is also installed onto an intermediary jig as an external setup procedure. During the setup, the next fixture is already attached and ready to be installed onto the machine. An example is standardizing the fixture subplate so that all fixtures are installed the same way. All fixtures are designed and build with an intermediary jig.

Step 3: Streamline the Setup Operation

In the final step of SMED, you can improve the remaining internal and external setup operations by evaluating each task's functions and purpose again. Methods for implementing improvements are separated into external and internal setup improvements. In developing the improvement plan, consider the seven forms of waste (*muda*):

1. Overproduction
2. Transportation
3. Inventory
4. Overprocessing
5. Waiting
6. Motion
7. Defects

(Refer back to Figure 8.1 in Chapter 8 for a description of each type of waste.) Your improvement plan should convert internal to external setup, eliminate or reduce both internal and external setup, and eliminate adjustments.

Streamlining External Setups

The external setup improvements focus on streamlining the storage and transportation aspect of setup. To refine these areas, tool and part management are key. 5S is essential to a successful changeover. 5S will ensure that you and your team does not lose time searching for tools and materials because the required materials will be in the proper place, and they will be clean and working. You can often cut setup times in half just by organizing your materials.

Streamlining Internal Setups

There are a few techniques for streamlining the internal setup:

■ Implementing parallel operations: Certain changeovers require tasks be performed at the front and back of the machine. Parallel operations reduce the time lost by walking back and forth from the front and back of the machine. Instead, divide the setup operations between two people, one for each side of the machine; this eliminates the walk time and reduces the internal setup time. However, you must carefully develop a detailed procedural chart to maintain safe and reliable operations during changeover. Your procedural chart should list the task sequence for each person, the time it will take, and when safety signals are required. The signal should be a buzzer, a whistle, or a light to clearly notify the other person.

■ Using functional clamps: Another technique to streamline internal setups is to use functional clamps. Bolts are considered an enemy in SMED because they slow down internal setups. Bolts often get lost, get mismatched, and take too long to tighten. In contrast, functional clamps attach items in place with a minimal amount of effort and can be loosened or tightened quickly. Also, because they are typically attached to the machine, operators cannot lose or mismatch them. Types of functional clamping systems include one-turn, one-motion, and interlocking systems.

■ Eliminating adjustments: Eliminating adjustments also reduces the time spent during internal setup. Trial runs and adjustments typically account for 50% of the total time in a changeover. By eliminating these adjustments, you avoid any time lost because of machine downtime. The key is to have the proper settings before starting the machine for the new operation. Trial runs and adjustments depend on the accuracy of centering, dimensioning, and condition setting. Eliminating adjustments can be achieved in several ways:

1. Using numerical scales and standardized settings; for example, you can make a graduated scale with marks that indicate the proper settings.

2. Identifying imaginary reference planes and centerlines; for example, by placing V-blocks and rods on the machine table parallel to the centerline and then aligning the center of the cutter.

3. Using the least common multiple (LCM) system, which takes into account operations that have elements in common but are different with respect to dimensions, patterns, or functions.

■ Mechanization: This is the final attempt to streamline setups. This is because mechanization will not significantly reduce the setup time as much as the other techniques. Another reason is that mechanization will reduce an inefficient operation, but it will not make the process better. Techniques in mechanization include:

1. Using forklifts for inserting large dies or molds into machines.

2. Moving heavy dies.

3. Tightening and loosening dies by remote control.

4. Using the energy from presses to move heavy dies.

Document the New Setup

The final step of SMED is to document the new setup on a new setup analysis chart. Videotape the setup procedure again and record the time for each element.

Develop a new improvement plan. This process will continue until the setup is eliminated or the setup is within takt time. The setup analysis chart is now the basis for the setup procedure because it contains the necessary steps for successful changeover. Any other detailed work instructions not included must also be posted and personnel trained. Remember that continuous improvement is a cycle, so now you should go back and try to streamline the setup process even more.

Conclusion

SMED enables significant reductions in changeover time, which also enables organizations to be more flexible for changes in customer demands and requirements. Now let us take a look at a technique for documenting processes. Chapter 14 discusses Standard Work in detail so that you can implement Standard Work in your organization.

Chapter 14

Standard Work: Documenting the Interaction between People and Their Environment

So far, we have covered three Lean tools: Six Sigma (described in Chapter 11), which focuses on improving product quality by reducing process variation; 5S (described in Chapter 12), which improves productivity by creating a clean and manageable workplace; and Single-Minute Exchange of Dies (SMED; described in Chapter 13), which improves productivity by reducing setup time. Standard Work is another Lean tool; it defines and documents the interaction between people and their environment.

Standard Work provides a routine for consistency of an operation and a basis for improvement by detailing the motion of the operator and the sequence of action. First, you document your current process to provide a basis or standard for continuous improvement. After you have seen improvements, then you should revise your department Standard Work to incorporate those improvements.

An Overview of Standard Work

Standard Work consists of three elements: takt time, Standard Work Sequence, and Standard Work-in-Process (which are discussed in detail in this chapter). Used as a tool, Standard Work accomplishes the following:

- It establishes a routine for repetitive tasks.
- It makes managing resource allocation and scheduling easier.
- It establishes a relationship between a person and the environment (both the machine and materials).

- It provides a basis for improvement by defining the normal process and high-lighting areas for improvement (by making problems visual and obvious).
- It prohibits backsliding.

The goal of the implementation project is to first record your existing process. Then, using time observations of your initial process, you should make changes to the line to improve working conditions, flow, and to level load of your operators. Finally, you should document your improved process to provide a baseline for your team to use in the future.

Standard Work is a tool to determine maximum performance with minimum waste through the best combination of operator and machine. The goal is to improve. Standard Work helps eliminate variability from the process. It functions as a diagnostic device. Standard Work also exposes problems and facilitates problem solving. It identifies waste and drives us to kaizen the process.

Prerequisites for Standard Work

To implement Standard Work, the operation must be observable, repetitive, and based on human motion. The process must be standardized with all variable processes kaizened. The floor supervisor must be responsible for the implementation of Standard Work.

Three Elements of Standard Work

The elements of Standard Work are takt time, Standard Work Sequence, and Standard Work-in-Process.

Takt Time

Takt time is how frequently a product must be completed to meet customer expectations. It is calculated using customer demand and available time. Takt time sets the rhythm for Standard Work.

Operator cycle time is the total time required for an operator to complete one cycle of operation, including time needed for walking, loading and unloading, and inspecting products. The *machine cycle time* is the time between the instant an operator presses the "on" or "start" button and the point at which the machine returns to its original position after completing the target operation.

Takt time is equal to the total daily operating time divided by the total daily requirements. The variables include customer demand and available work time. Therefore, you should recalculate takt time when your customer demand changes or your available work times change.

$$\text{Takt time} = \frac{\text{Available time}}{\text{Customer demand}}$$

Standard Work Sequence

Standard Work Sequence is the specific order in which an operator performs the manual steps of a process. The Standard Work Sequence may be different from the *process sequence*. Focusing on the Standard Work Sequence identifies waste and stabilizes the process. Standard Work Sequence requires multiskilled operators. A complete operator cycle is from the time the operator begins the sequence to the time the operator returns to that same point.

Standard Work-in-Process

Standard Work-in-Process is the minimum amount of parts on the line that will allow an operator to flow product efficiently. Keeping the number of parts standard is key to allowing work to continue without the operator waiting.

Documenting Standard Work

Several standard forms are used for documenting Standard Work. These forms help standardize the process and are described in detail in this section.

Takt Time Sheet

The takt time sheet documents all of the following information:

- Minutes available
- Pieces per shift
- Allotted time for break times
- Allotted time for wash times
- Allotted time for cleanup

Figure 14.1 shows the takt time sheet. In Chapter 7, a completed example of this form is provided for Carjo in Figure 7.1.

Time Observation Sheet

The time observation sheet focuses on manual and walk time elements. You should fill out a separate sheet for each operator. The time observation sheet has three key steps:

1. Identify work elements.
2. Determine the observation points.
3. Time each element with a running clock.

The first step in appraising the current process is to time the process. Begin by outlining the process steps. Then time the existing process for walking, loading, unloading, standard inspection, and machine cycle times. You should take ten observations of the process and record them on your time observation form (refer to Figure 14.2).

When making your observations, you can use a running clock. In this case, you should subtract the times between observations to obtain the time for each step. Then determine the lowest elemental time for each step and then add the

Calculating takt time
_____ Hours =   Minutes (based on standard work shift)
Minutes (break time)
Minutes (wash time)
Minutes (clean-up)
Minutes (team meetings)
Total   _____   Available minutes per shift
_____ Minutes available x 60 = _____ Seconds per shift
_____ Seconds per shift x 2 shifts = _____ Seconds per day
_____ Seconds divided by _____ pcs/day = _____ Seconds
Takt time = _____ Seconds per piece

Figure 14.1   Takt time sheet.

Process for observation	Time observation form										Part no.		Part type	
											Part name		Daily demand	
No.	Component task	1	2	3	4	5	6	7	8	9	10	Low elem. time	Adj.	Adj. elem. time
1														
2														
3														
4														
5														
6														
7														
8														
9														
10														
Time for one cycle														

Figure 14.2   Time observation form—blank template.

lowest elemental times to obtain the time for one cycle. You should also add together the times for individual steps to calculate the time for one cycle. You can then make adjustments to the steps to make the total of the lowest elemental times equal to the lowest cycle time of your actual observations.

If the total time for one cycle of your actual observations is greater than the takt time for the operator performing the machining responsibilities, you will know that your process must be improved to meet takt time. Figure 14.2 provides the time observation sheet and Figure 14.3 is an example of a completed time observation sheet.

Process Capacity Table

The process capacity table documents the machine capacity per shift. Use one sheet for each cell. The table focuses on the total machine time, which includes any load and unload times. Document only the load, unload, and cycle start when calculating the manual time. You should also take into account tool changes in your calculations. Do not include any abnormalities that are not standard.

You should perform process capacity calculations for each process step. List each process step with its associated manual time, machine time, and walking time. In addition, evaluate tool changes. Then add together the manual time, machine time, and walking time to obtain the total time to complete the process step. Finally, divide this time into the available operating time per shift; this calculation results in the processing capacity per shift. Figure 14.4 provides the process capacity table, and Figure 14.5 is an example of a completed process capacity table.

Standard Work Combination Sheet

The next step of outlining the existing process is to fill out the Standard Work Combination Sheet. The Standard Work Combination Sheet combines manual, automatic machine, and walk elements. Plot these against the takt time. Use one sheet for each operator, and you should post the sheet at the starting point of each operator sequence. Figure 14.6 shows the Standard Work Combination Sheet and Figure 14.7 provides an example.

Fill out a Standard Work Combination Sheet for each operator. Use the same steps to complete this form as you used when completing your process capacity sheet. Then list the manual, machine, and walking times for their respective process step.

■ Drawing a straight line first, draw the manual time on the Combination Sheet chart.
■ Then add the machine cycle time from the manual time by drawing a dotted line.
■ Finally, draw a curved line from the end of the manual time for the observed length of walking time down to the next process step.

No.	Process for observation	Time observation form										Part no.		Part type	
												Part name		Daily demand	
	Component task	1	2	3	4	5	6	7	8	9	10	Low elem. time	Adj.	Adj. elem. time
1	Unload, blow off chips, load, clamp, start	:33	:37	:38	:43	:40	:43	:56	:72	:57	:52	:33	:10	:43
		:33	:37	:38	:43	:40	:43	:56	:72	:57	:52			
	Walk	:10	:11	:10	:10	:11	:09	:19	:12	:11	:15	:09	:00	:09
		:43	:48	:48	:53	:51	:52	1:15	1:24	1:08	1:07			
2	Unclamp, unload, blow off chips, load, clamp, start	:32	:38	:37	:41	:30	:36	:36	:33	1:41	:41	:30	:10	:40
		1:15	1:26	1:25	1:34	1:21	1:28	1:51	1:57	1:49	1:48			
	Walk	:18	:16	:14	:14	:17	:17	:24	:19	:17	:17	:14	:00	:14
		1:33	1:42	1:39	1:48	1:38	1:45	2:15	2:16	2:06	2:05			
3	Unload, blow off chips, load, start	:27	:35	:32	:29	:23	:29	:29	:34	:40	:43	:23	:05	:28
		2:00	2:17	2:11	2:17	2:01	2:14	2:44	2:50	2:46	2:48			
	Walk	:10	:08	:07	:08	:09	:09	:09	:08	:11	:07	:07	:00	:07
		2:10	2:25	2:18	2:25	2:10	2:23	2:53	2:58	2:57	2:55			
4	Unload, blow off chips, load, start	:24	:33	:28	:28	:30	:29	:35	:38	:44	:43	:24	:05	:29
		2:34	2:58	2:46	2:53	2:40	2:52	3:27	3:36	3:41	3:38			
	Walk	:08	:12	:08	:11	:09	:10	:12	:09	:09	:09	:08	:00	:08
		2:42	3:10	2:54	3:04	2:49	3:02	3:39	3:45	3:50	3:37			
5	Unload, blow off chips, load, start	:31	:30	:41	:33	:32	:33	:29	:33	:36	:40	:29	:05	:34
		3:13	3:40	3:35	3:37	3:21	3:35	4:08	4:18	4:26	4:17			
	Inspect	:13	:13	:14	:13	:13	:14	:13	:14	:14	:13	:13	:00	:13
		3:26	3:53	3:49	3:50	3:34	3:49	4:21	4:32	4:40	4:30			
	Walk	:08	:08	:07	:08	:07	:07	:07	:08	:07	:07	:08	:00	:08
		3:34	4:01	3:56	3:58	3:41	3:56	4:28	4:40	4:47	4:37			

Figure 14.3   Example time observation sheet.

#	Task													
6	Unload, blow off chips, load, start	:31	:30	:30	:32	:33	:34	:34	:36	:29	:49	:29	:05	:34
		4:05	4:31	4:26	4:30	4:14	4:30	5:02	5:16	5:16	5:26			
	Inspect	:16	:13	:19	:14	:13	:15	:13	:16	:14	:14	:13	:00	:13
		4:21	4:44	4:45	4:44	4:27	4:45	5:15	5:32	5:30	5:40			
	Walk	:03	:06	:04	:04	:03	:05	:03	:04	:04	:03	:03	:00	:03
		4:24	4:50	4:49	4:48	4:30	4:50	5:18	5:36	5:34	5:43			
7	Unload, blow off chips, load, start	:33	:29	:26	:29	:30	:31	:33	:35	:28	:30	:28	:05	:33
		4:57	5:19	5:15	5:17	5:00	5:21	5:51	6:11	6:02	6:13			
	Inspect	:08	:12	:10	:09	:08	:10	:09	:09	:08	:09	:13	:00	:13
		5:05	5:31	5:25	5:26	5:08	5:31	6:00	6:20	6:10	6:22			
	Walk	:12	:08	:10	:08	:08	:09	:08	:10	:09	:09	:08	:00	:08
		5:17	5:39	5:35	5:34	5:16	5:40	6:08	6:30	6:19	6:31			
8	Unload, load, blow off chips, start	:33	:20	:19	:23	:20	:25	:25	:27	:27	:32	:19	:05	:24
		5:49	5:59	5:54	5:57	5:36	6:05	6:33	6:57	6:46	7:03			
	Walk	:05	:05	:05	:07	:07	:06	:05	:06	:06	:15	:05	:00	:05
		5:54	6:04	5:59	6:04	5:43	6:11	6:38	7:03	6:52	7:18			
9	Blow out, put gloves on	:57	:63	:67	:65	:63	:65	:66	:64	:65	:67	:57	:00	:57
		6:51	7:07	7:06	7:09	6:46	7:16	7:44	8:07	7:57	8:25			
	Plug, blow off chips, remove gloves	:78	:85	:89	:80	:79	:85	:84	:86	:82	:85	:78	:00	:78
		8:09	8:32	8:35	8:29	8:05	8:41	9:08	9:33	9:19	8:50			
	Walk	:20	:13	:09	:17	:22	:29	:32	:28	:30	:29	:09	:00	:09
		8:29	8:45	8:44	8:46	8:27	9:10	9:40	10:01	9:49	9:19			
	Time for one cycle	8:29	8:45	8:44	8:46	8:27	9:10	9:40	10:01	9:49	9:19	7:40	:50	8:30

Figure 14.3 Example time observation sheet (Continued).

After you have determined whether the current process is within takt time, evaluate the process to identify areas for kaizen.

Standard Worksheet

After documenting your process capacity and your time observations, the next step is to draw a Standard Worksheet to depict the process flow and machine

Department manager			Process capacity form					Part no.			Part type		Operating time per shift in seconds
Supervisor								Part name			Daily demand		
Step no.	Process description	Machine no.	Base time (seconds)		Tool change				Time (seconds)			Processing capacity	Remarks
			Manual	Machine	# of pcs per change	Replacement time	Tool change time		Replacement time	Total time to complete			
1													
2													
3													
4													
5													
6													
7													
8													
9													
10													
	Total												

Figure 14.4 Process capacity table—Blank template.

Department manager		Process capacity form							Part no.	Part name	Part type		Operating time per shift in seconds
Supervisor											55		26,100
Step no.	Process description	Machine no.	Base time (seconds)		Tool change			Time (seconds)		Daily demand	Processing capacity		Remarks
			Manual	Machine	# of pcs per change	Replacement time	Tool change time		Total time to complete				
1	Machine Z-plane	CNC 1	53	294	800	30	360		347		75		
2	Drill bolt holes	CNC 2	33	414	1500	30	90		447		58		Bottleneck operation
3	Rough drill tube bores	CNC 3	28	349	750	30	90		377		69		
4	Finish drill tube bores	CNC 4	27	196	750	30	90		223		117		
5	Rough and finish drill center bore	CNC 5	32	423	375	30	90		455		57		Bottleneck operation
6	Mill clearance cut	CNC 6	34	378	750	30	360		413		63		
7	Drill oil hole	CNC 7	30	343	2500	30	60		373		70		
8	Hand tap	CNC 7	22	229	75	90	0		252		103		
9	Assemble gear housing	Assy.	59	0	0	0	0		59		442		
		Total	316	2,626									

Figure 14.5 Example process capacity table.

Figure 14.6   Standard Work Combination Sheet—Blank template.

Standard Work Combination Sheet

Model								Date prepared		Quota per shift	55	Manual work	———
Work seq.								Group		Takt time	474	Machine work	– – –
												Walking	⌇⌇⌇

Step no.	Description of operation	Time			Operating working time (seconds)
		Manual	Auto	Walk	
1	Unload, blow off chips, load, clamp start	53	294	9	
2	Unclamp, unload, blow off chips, load, start	33	414	15	
3	Unload, blow off chips, load, start	28	349	7	
4	Unload, blow off chips, load, start	27	196	13	
5	Unload, blow off chips, load, start	32	423	21	
6	Unload, blow off chips, load, start	34	378	19	
7	Unload, blow off chips, load, start	30	343	20	
8	Unload, blow off chips, load, start	22	229	7	
9	Blowout, plug, blow off chips, unload	59	0	19	
10					
Total		318	X	130	

Figure 14.7   Example Standard Work Combination Sheet.

layout of the current process. The Standard Worksheet is an overhead view of the cell or operations that illustrates the process and Standard Work Sequence. It documents the Standard Work-in-Process, safety precautions, and quality checks. Again, you should complete a sheet for each operator, and you should post the Standard Worksheet at the starting point of each operator sequence. Figure 14.8 shows the Standard Worksheet and Figure 14.9 shows an example.

Draw the operation sequence from raw material to finished material. First, draw the machine and assembly stations to illustrate your floor layout. Then add the operator flow to show the steps the machining and assembly operator(s) take(s) to complete the process. You should number these steps in sequence for each operator. Then mark each station for quality checks, safety precautions, and work-in-process.

Operator-Loading Chart

The final charting of the Standard Work process is to graph the operator loading chart. The operator loading chart documents the time allocated for all operators in the cell. It also documents how many operators are in the cell. The chart is a bar chart. Figure 14.10 shows an example, with the graph depicting the operator's cycle time(s) versus their respective takt times. It shows if the operators are able to do their assigned tasks within takt time. If the operators' time is more than the takt time, then they will not meet customer demand. However, if the operator time is significantly less than takt time, this is an indication that there is waste in the process (i.e., either the operator or the machine is waiting or idle). For example, in Figure 14.10, Operator A has a cycle time of 460 s, and Operator B has a cycle time of 458 s. The takt time is 474 s, which indicates both operators can complete their respective tasks within takt time.

Least-Operator Concept

Because the purpose of Lean is to drive continuous improvement, there is a technique within Standard Work that helps drive continuous improvement. The least-operator concept states that the cell should be frontloaded, and you should allocate all waiting time to the least operator. The least operator should be the last operator in the sequence. The other operators should be fully loaded to takt time. This makes the waste and the waiting time visible. It also exposes the opportunity for improvement at the last operation.

In contrast, natural work has the operator at various loads. The traditional work setup has the operators at equal loads; however, this is usually not at takt time. The frontloaded concept has all of the operators except the last operator loaded at takt time.

Operation	From:				**Standard Worksheet**				Part no.				
Sequence	To:								Part name:				
		Quality check		Safety		Standard WIP	# Pieces WIP	Takt time		Cycle time			
		◇		●									

Figure 14.8   Standard Worksheet—Blank template.

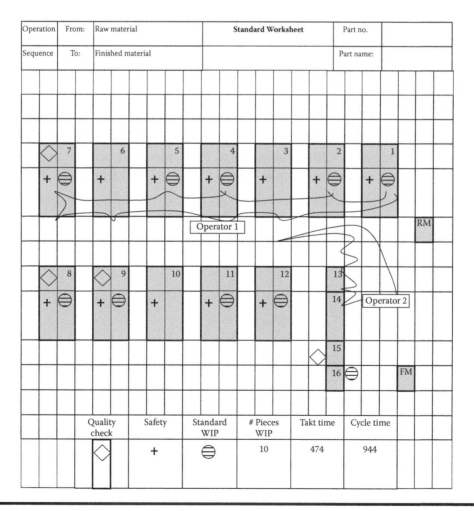

Figure 14.9   Example Standard Worksheet.

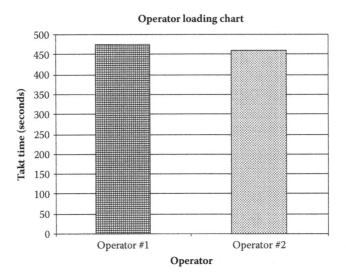

Figure 14.10 Operator loading chart.

Resource requirements can be calculated. The total number of operators required is equal to the sum of individual operator cycle times divided by the takt time.

Conclusion

Standard Work documents a process so that all operators follow the same process to reduce process variability. Now let us take a look at how to improve quality. Chapter 15 takes you through mistake-proofing devices to ensure defects are not produced.

Mistake-Proofing (a.k.a. Poka-Yoke): Preventing Defects by Monitoring Process Conditions and Correcting Errors at the Source

The final tool is poka-yoke, a methodology that focuses on preventing defects and improving quality. The goal of zero defects is achievable. The Zero Quality Control system approach of mistake-proofing (*poka-yoke*, in Japanese) prevents defects by monitoring process conditions and correcting errors at the source. It is human nature to make mistakes. In this approach, poka-yoke devices are used to perform 100% inspections and give feedback about each part or operation. Installing a poka-yoke device has a considerable effect on quality, but it can also give you a false impression that these devices alone will eliminate defects. However, if you combine poka-yoke systems with self-checks or successive checks, you can effectively obtain 100% inspections, prompt feedback, and action. To do that, it is essential for you to understand how to use poka-yoke systems, their functions, the types of systems, and various detection methods—all of which are covered in this chapter.

An Overview of Poka-Yoke/Mistake-Proofing

Poka-yoke is a technique to prevent simple human error. Preventing the defect before it is produced is, of course, the most effective means of reducing defects. However, implementing poka-yoke devices to detect errors and immediately stop the action is also a valuable part of the continuous improvement effort.

The concept of poka-yoke has existed for a long time in various forms. Shigeo Shingo, a Japanese manufacturing engineer, developed the idea into a tool to achieve zero defects. The idea behind poka-yoke is focused on taking over repetitive tasks and actions to free the worker's time and mind for more creative and value-adding activities.

Zero Quality Control is an approach for achieving zero defects. It is based on the idea that controlling process performance can prevent defects even when a machine or a person makes mistakes. Zero Quality Control is a blameless approach that recognizes that people sometimes make mistakes.

Focusing on producing zero-defect products is essential to maintaining customer satisfaction and loyalty. Cost is another reason to focus on eliminating defects. Defects result in costs from scrapping a product, reworking, or repairing damaged equipment. Zero defects are also key for a company to achieve Lean production and smaller inventories. Reducing defects allows a company to decrease buffer inventories that are built with anticipation of problems. Companies are able to produce the exact quantity of products ordered by the customer.

Inspection Techniques

Three major inspection techniques exist in the field of quality control:

1. Judgment inspection: With this technique, the operator separates defective parts from good parts after processing. This method of inspection prevents defects from reaching the customer; however, it does not lower the company's internal defect rate.
2. Informative inspection: With this technique, you investigate the cause of the defect and relay that information back to the appropriate process so that the operator can take action to reduce the defect rate.
3. Source inspection: Because defects are typically caused by simple mistakes, this approach to inspection focuses completely on the source so that the mistake can be corrected before it even becomes a defect. Zero defects can be achieved using source inspection.

Zero Quality Control consists of three main methods leading to eliminating defects. First, source inspection checks for the factors that *cause* an error, rather than inspecting for the *resulting defect*. It assures that certain conditions exist in order to perform a process properly. An example of source inspection is adding a locator pin to prevent a part from being misaligned in a fixture. Source inspection differs from judgment inspection and informative inspection in that it catches errors and provides feedback so that the errors can be corrected before processing the product.

Secondly, inexpensive poka-yoke devices 100% inspect automatically for errors or defective operating conditions. Zero Quality Control varies from statistical quality control inspections in that it inspects every single product produced.

Statistical quality control only gives an idea of whether a process is in control and does not prevent defects. A limit switch or inexpensive sensing device is an example of a 100% inspection device.

The third component is taking immediate action. The operations are stopped immediately when a defect or mistake is made and will not resume until the mistake is corrected. An interlocked circuit that automatically shuts down a machine when a mistake is made is an example of taking immediate action.

Types of Errors

There are several different types of errors:

1. The first type of human error is *forgetfulness*, which can occur when an operator is not concentrating. A safeguard to prevent forgetfulness is setting inspections for the operators to perform at regular intervals.
2. *Errors due to misunderstanding* can happen when people make conclusions before they are familiar with the situation. Training and standardizing work procedures can help avoid these situations.
3. *Identification errors* are another type of human error. Situations can be misjudged when viewed too quickly or from too far away to be clearly visible. Training and attentiveness can avoid these.
4. *Lack of experience* can also cause newer operators to make errors. Skill building can help avoid these types of mistakes.
5. The fifth type of human error is *willful errors*, when operators decide they can ignore certain rules. Experience and basic education are safeguards against these errors.
6. *Inadvertent errors* can also be made when people are absentminded or lost in thought. Through proper discipline and work standardization, these defects can be avoided.
7. *Delays in judgment* can also cause errors. Again, skill building and work standardization are aspects to avoiding these defects.
8. *Lack of suitable work standards or instructions* can cause mistakes. Work instructions and work standardization can help avoid these errors.
9. Occasionally, equipment and machines will run differently than expected, resulting in *surprise errors*. Using Total Productive Maintenance (TPM) can avoid these errors.

Types of Defects

There are various types of defects, including:

1. Omitted processing (i.e., a missed process).
2. Processing errors (i.e., a broken tool).

3. Errors in setup (i.e., the wrong machine setting).
4. Missing parts (i.e., a missing component in an assembly).
5. Wrong parts (i.e., assembling the wrong component).
6. Processing the wrong part (i.e., picking up the wrong part).
7. Misoperation (i.e., variation in the process).
8. Adjustment errors (i.e., changing the setting incorrectly).

Defects typically occur during one of five situations:

1. Defects can occur because of inappropriate procedures or standards during process planning. Proper planning to ensure correct standards can avoid this situation.
2. Defects can also occur because of excessive variability in a process. Maintenance can prevent these types of defects.
3. Defects can occur when material is damaged or varies excessively. A means of eliminating this situation is implementing inspection on receipt of the materials for defects and variation.
4. Worn equipment and tools can cause defects. Again, regular maintenance can prevent these defects.
5. The final situation can occur even when the above situations do not exist: simple human mistakes occur that result in the production of defective products.

Defects either are about to occur or already exist. Poka-yokes have three main functions: shutdown, control, and warning. *Prediction* is recognizing that a defect is going to happen. *Detection* is recognizing that a defect has occurred.

Quality Improvement Activities

The traditional quality improvement cycle consists of "Plan-Do-Check," as shown in Figure 15.1. The processing conditions are determined in the Plan stage. These planned actions then occur in the Do stage. Finally, the Check stage performs the quality monitoring, in which information regarding defects is relayed back so that you can take corrective action in the next Plan stage in order to improve conditions during the next Do stage.

However, you cannot completely prevent defects even when continuously repeating the functions in this cycle because feedback about defects is given only after a defect occurs. There is no means of preventing the error before it happens.

Zero Quality Control addresses this problem by integrating the Do and Check stages, as shown in Figure 15.2. Errors can occur between the Plan and Do functions. In the Zero Quality Control approach, inspection is carried out at the point where the error happens. When an error is detected, the operator is able to correct the problem before the work is done, using source inspection.

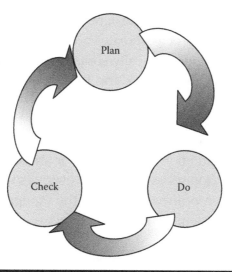

Figure 15.1   Plan-Do-Check cycle of traditional quality improvement cycle.

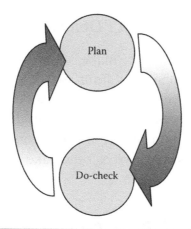

Figure 15.2   Integrated Do and Check in the Zero Quality Control approach.

Poka-Yoke Detection Systems

There are several methods to identify the proper poka-yoke for a situation. On the basis of the situation, there are appropriate systems to prevent defects.

■ The first method is to identify an item by its characteristic. This can be the weight, dimension, or shape of the product.
■ Another method is to determine how the defect can deviate from procedures or omitted processes by following the process sequence.
■ A third method is detecting deviations from fixed values. This can be achieved by using a counter, odd-part-out method, or critical condition detection.

In Zero Quality Control, true poka-yokes are used as source inspections to detect errors before the production process creates a defective product. The poka-yoke system detects an error and automatically shuts down the equipment or gives a warning. Control systems are used to stop the equipment when an error is detected. This is a more effective means of achieving zero defects because it does not depend on the operator. This type of system may not always be possible or convenient; therefore, a warning system may also be used. A warning system is used to get the attention of the operator and could be in the form of a flashing light or a sound. Control systems, however, also typically use lights and noise to direct the attention of the operator to the equipment with the problem. Non-automated warning systems can also be effective, such as color-coding parts and part holders.

Poka-yoke detection devices can be categorized into three main methods: contact methods, fixed-value methods, and motion-step methods. These methods can be used with either control systems or warning systems, but each uses a different approach to defects or errors.

Using Contact Methods to Detect Defects or Errors

Contact methods detect if a product makes a physical or energy contact with a sensing device. Micro switches and limit switches are the most commonly used contact devices. However, contact methods do not have to be very technical. Inexpensive contact devices such as guide pins or blocks will not allow a part to be loaded into a fixture in the wrong position. These types of contact methods are "passive devices" and take advantage of a parts design or uneven shape.

Using Fixed-Value Methods to Detect Defects or Errors

You should use a fixed-value poka-yoke method to detect defects or errors when you have a process with a fixed number of parts that must be attached or assembled in a product. You can also use this method when you have a process that requires a fixed number of repeated operations to be performed at a station.

With fixed-value methods, you use a device to count the number of times a task is done; the device then signals or releases the part only when the required number is reached. Limit switches can be used with each movement sending a signal to the counter that, in turn, detects when the required number of movements is complete. The device will only release the part when the preset number of signals is reached. For example, if you have an assembly with three bolts that must be set at a specific torque, the assembly fixture would hold the part in place until the count (in this case, three) was reached, and the device would be preprogrammed such that it would only count up when the torque level is reached.

Using Motion-Step Methods to Detect Defects or Errors

Another approach is to use the motion-step method to sense whether a motion has been completed within a certain expected time. You can also use this method to detect whether tasks are performed according to a specified sequence, which is often a helpful tool for assembling the proper parts in a particular product model.

Conclusion

Mistake-proofing improves the overall quality of a product or service. Now that you have become more familiar with the Lean Six Sigma toolbox, let us take a look at how you can use these tools to drive improvements that are tied to the strategic vision. The next part of the book shows you how to cascade down the strategic vision and goals into the daily activities of your organization. The daily activities utilize the tools that were discussed in this part of the book to realize significant gains.

PLAN AND IMPLEMENT   V

The final part of this book enables you to cascade down the strategic goals of the organization to implement process improvements. This enables you to develop action plans to integrate the strategic goals into the daily activities of everyone in the organization.

Chapter 16

Case Study: Daily Management and Action Plans at Carjo Manufacturing Co.

Now that you have a better understanding of the Lean Six Sigma tools described in Chapters 11–15, let us revisit Carjo Manufacturing to see how it drives down the strategic goals into its daily management and action plans using Hoshin Kanri. Recall (from Chapter 4) that the Hoshin action plan further drives down the strategic goals of the organization to targets and milestones for each implementation strategy. A Hoshin action plan should be created for each implementation strategy.

Developing Carjo's Hoshin Action Plan

Recall from Chapter 5 that the leadership team at Carjo selected the strategic goal of implementing cost-reduction projects because of this goal's impact on safety, quality, delivery, and cost. The department managers reviewed their current-state value stream maps (refer back to Figures 8.2, 8.3, and 8.4) with their teams to identify improvement opportunities for cost-reduction projects. Also, the manager of the tube line noted a high internal defect rate in the tube line for oversize bushing bores (refer to Figure 8.2).

The oversize bushing bores currently account for 4.3% of the total product costs. This background information is key to understanding and developing the situation summary for Carjo's Hoshin action plan. There should be a compelling reason that is linked to the strategic objective and based on quantitative data. Without the clear linkage, you may be focusing on improvements that do not impact the overall goals of your organization and therefore may be suboptimal improvements.

In addition, the defect had a current internal quality of 56,704 parts per million (PPM). The department manager, Paul Farber, selected this project as part of the company's cost-reduction strategy. Using their current baseline, the team developed short- and long-term goals for the project.

The team had been working for several months to reduce the defect rate, but it was still not realizing significant gains. The improvement team could not quantify the key variables that affect the product performance. Therefore, the team decided to pursue this defect issue as a Six Sigma project.

The targets and milestones developed for this project follow the Six Sigma methodology. In this example, the team has outlined its target actions as conducting a measurement systems analysis (MSA) and a process failure modes and effects analysis (PFMEA) with expected completion dates. Figure 16.1 shows the Hoshin action plan developed by the team.

Hoshin action plan		
Core objective: Implement cost reduction projects	**Team:** Kayla M., Zoe W., Derek C.	
Management owner: Paul Farber	**Date:** 9/14	
Department: Tube line	**Next Review:** 10/15	
Situation summary: Internal defects are currently 4.3% of total product cost		
Objectives(s): Implement cost reduction projects to improve financial returns		
Short-term goal: Internal quality to 32,798 PPM	**Strategy:** Six Sigma project on oversize bushing bore	**Targets and milestones:** Conduct measurement systems analysis 9/21
Long-term goal: Internal quality to 28,154 PPM		Process failure models and effects analysis 10/2

Figure 16.1　Carjo Hoshin action plan.

The department as a team goes through each core objective to identify implementation strategies that will impact the overall organization. The team develops a Hoshin action plan for each of these strategies.

Developing Carjo's Hoshin Implementation Plan

Next, each of these strategies is cascaded down to the Hoshin implementation plan, as shown in Figure 16.2. The team at Carjo lists each implementation strategy in the left-hand column. In order to determine the target improvements, the team revisits the current-state maps for this process. The current-state map provides the baseline for the improvement strategy. By understanding the current performance level, the team determines the target improvement.

Once the team members determine their target improvement, they break down this improvement into monthly improvement targets. This enables the team to manage the project and monitor trends. Following the example of the Six Sigma project to reduce oversize bushing bores, the team breaks down the improvement in internal PPM by month. The team can then monitor if they meet their targeted PPM reduction each month.

The team should present this chart to its senior leadership team on a monthly basis. It provides a high-level team picture of current improvement activities. The team can also highlight the projects that are on track by coloring the background green for "project on track" and red for "projects that are not meeting the specified target." In a quick glance, anyone can then see the status. However, because this book is in black and white only, this color scheme does not work as well. Therefore, in Figure 16.2 the projects that are not meeting the set target are shaded in grey, and projects that are on track have no background fill.

Preparing Carjo's Hoshin Implementation Review

The final form is the Hoshin implementation review, shown in Figure 16.3. This is where the team tracks its performance status, implementation issues, and performance measurement. Each implementation strategy is carried down from the Hoshin implementation plan. A Hoshin implementation review form should be created for each implementation strategy.

For the Six Sigma project to reduce oversize bushing bores, the team has outlined three upcoming steps in their project (see "Performance Status" on Figure 16.3). These include:

1. Conduct an MSA.
2. Conduct a PFMEA.
3. Conduct design failure modes and effects analysis (DFMEA).

Hoshin implementation plan

Core objective:
Implement cost reduction projects

Strategy owner:
Miya F.

Date:
9/14

Strategy	Performance		Schedule and milestones											
			Jan	Feb	Mar	Apr	May	Jun	Jul	Aug	Sep	Oct	Nov	Dec
Six Sigma project on oversize bushing bore	Target	28,154 PPM	57,304	54,654	52,004	49,354	46,704	44,054	41,404	38,754	36,104	33,454	30,804	28,154
	Actual	31,652 PPM	54,007	54,900	51,583	48,128	42,860	43,957	38,100	37,905				
Kaizen participation	Target	100%	10%	20%	30%	40%	50%	60%	70%	80%	90%	100%	100%	100%
	Actual	70%	12%	23%	31%	42%	55%	58%	74%	82%				

Figure 16.2   Carjo Hoshin implementation plan.

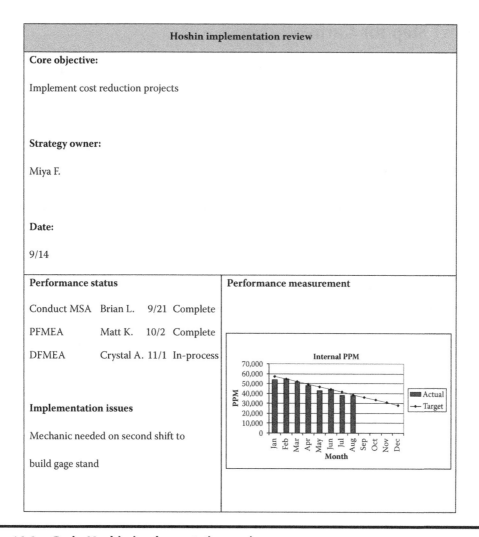

Figure 16.3   Carjo Hoshin implementation review.

The team has assigned ownership to an individual team member with an expected completion date. A status is also provided by the team as to whether the item is complete, in-process, or scheduled.

The next important items are the implementation issues. As part of the MSA (see the bottom left of Figure 16.3), the team identified the need for a gage stand. Here, the team also highlighted their trouble with securing a second shift mechanic to build the gage stand. By presenting this information, the leadership team can secure the necessary resources to make the team successful.

Finally, the team shows their performance trend (on the right-hand side of Figure 16.3). Because the goal is to reduce internal PPM for the bushing bore, the team shows a bar chart of the internal PPM trend.

The Hoshin implementation reviews should also be posted in the department. The team then can have a stand-up weekly meeting to review the status of the projects. This will help ensure that the projects are on track before the next monthly review with the senior leadership team.

The Next Step for Carjo

The next crucial step for Carjo is to extend the technique outlined in this book throughout the supply chain. After all, a company is only as good as its suppliers. To develop partnerships and ensure long-term success, the leaders at Carjo should take these methods to their suppliers once they have their internal strategy functioning and engrained into their daily habits. The methodology would use the macro value stream map to show the wastes in the supply chain.

Macro value stream mapping extends beyond the plant-level maps. Macro mapping can be done after creating current- and future-state maps inside the facility. These maps are created because a large portion of costs consists of purchased materials, and downstream inconsistencies can affect your facility's leanness. Also, added costs downstream can negate the cost savings you achieve internally. This can affect whether or not Carjo can increase their sales. The whole picture allows you to identify major asset reconfigurations by showing who does what where and with what tools.

The facility closest to the customer should be mapped first. All of the following information should be collected:

- Frequency
- Distance
- Cost
- Processing time
- Lead time
- Inventories
- Cost per unit
- Daily volume
- Shift data variation
- Frequency variation
- Demand variation

In the ideal state of macro value stream mapping, all activities are located in the exact process sequence.

Conclusion

Integrating Hoshin Kanri into daily activities is the final step in becoming a strategic Lean organization. This creates a linkage of common goals throughout all levels of the organization.

Conclusion

Lean is a philosophy that must be fully embraced by an organization to truly reap the full potential. Unless you tie Lean into the strategic vision of your organization, your employees may view it as merely a "flavor-of-the-month" strategy, which, of course, means they will not embrace it, implement it, or even take it seriously. In addition, Lean projects that are selected based on their impact on the entire organization will have the most effective results.

You must ensure that your strategic vision cascades down throughout your organization into the daily activities of everyone. This clear linkage enables an organization to move in a common direction with common goals. Hoshin Kanri, also known as Policy Deployment, enables the strategic vision to be carried down throughout the organization.

When employees understand the direction of the organization, they can make the appropriate improvements that will enable long-term success. By using strategic vision, your organization can employ Lean techniques to eliminate waste and improve flow.

Achieving your envisioned future-state value stream through Hoshin Kanri provides the principles and key tools to guide you in strategic Lean initiatives. As you target improvement opportunities, you will be able to see how they impact your organization's overall strategic vision. The alignment within your organization will also become clear.

LEANcyclopedia Lean Glossary

Andon: A line indicator light or board hung above the production line to act as a visual control. Andons are used to visually signal an abnormal condition.

Autonomation: Autonomation with a human touch or transferring human intelligence to a machine. This allows the machine to detect abnormalities or defects and stop the process when they are detected. Also known as *Jidoka*.

Bar chart: A graphical method that depicts how data fall into different categories.

Batch-and-queue: A mass production practice of producing lots and sending the batch to wait in a queue before the next operation.

Benchmarking: An activity to establish internal expectations for excellence, on the basis of direct comparison to "best." In some cases, the best is not a direct competitor in your industry.

Black Belt: A person trained to execute critical projects for breakthrough improvements to enhance the bottom line.

Brainstorming: A technique to generate a large number of ideas in a short period of time.

Breakthrough objectives: In Policy Deployment, those objectives characterized by multifunctional teamwork, significant change in the organization, significant competitive advantage, and major stretch for the organization.

Cell: A logical and efficient grouping of machines or processes that enables one-piece flow.

Cellularization: Grouping machines or processes that are connected by work sequence in a pattern that support flow production.

Cellular manufacturing: Manufacturing with the use of cells. See *cell*.

Chaku-Chaku: Japanese term for "Load-Load." It refers to a production line raised to a level of efficiency that allows the operator to simply load the part and move on to the next operation. No effort is expended on unloading. See *Hanedashi*.

Champion: An individual who acts as the sponsor or owner of a project and has the authority and responsibility to inform, support, and direct a team. Typically the individual is a director or vice-president-level manager. Also known as a mentor or sponsor.

Changeover: Altering a process to manufacturing a different product.

Changeover time: As used in manufacturing, the time from when the last "good" piece comes off of a machine until the first "good" piece of the next product is made on that machine. Includes warmup, first-piece inspection, and adjustments. Changeover times can be reduced through the use of SMED.

Cost of Poor Quality (COPQ): Costs associated with not doing things right the first time. Examples of COPQ include scrap, rework, and waste.

Countermeasures: Immediate actions taken to bring performance that is tracking below expectations into the proper trend. Requires root-cause analysis.

Curtain effect: A method that permits the uninterrupted flow of production regardless of external process location or cycle time. Normally used when product must leave the cell for processing through equipment that cannot be put into the cell (i.e., heat-treat, painting). Curtain quantities are calculated using the following formula: curtain quantity = per unit cycle time of curtain process/takt time.

Customer: Anyone who uses or consumes a product or service. A customer can be internal or external to the provider.

Cycle time: The time from the beginning of one operation in a process until it is complete.

Defect: A nonconformance in a product or service.

Design for manufacture and assembly (DFMA): A philosophy that strives to improve costs and employee safety by simplifying the manufacturing and assembly process through product design.

Deviation: The difference between an observed value and the mean of all observed values.

Downtime: Lost manufacturing time due to equipment, material, information, or manpower.

Economic value added (EVA): A residual income measure that subtracts the cost of capital from the net operating profits after taxes (NOPAT). It is the financial performance measure most closely linked to shareholder value and the cornerstone for a financial management and incentive compensation system that makes managers think and act like owners.

Failure mode and effects analysis (FMEA): A structured approach to assess the magnitude of potential failures and identify the sources of each potential failure. Corrective actions are then identified and implemented to prevent failure occurrence.

5S: A method of creating a self-sustaining culture that perpetuates an organized, clean, and efficient work place. Also referred to as the five pillars of the visual workplace.

Five-Why method: A simple problem-solving method of analyzing a problem or issue by asking "Why?" five times. The root cause should become evident by continuing to ask why a situation exists.

Fixed costs: Costs of production that do not change when the rate of output is altered.

Flexibility: The ability to respond to changes in demand, customer requirements, etc.

Flowchart: A pictorial representation of a process that illustrates the inputs, main steps, branches, and outcomes of a process. A problem-solving tool that illustrates a process. It can show the "as is" process or "should be" process for comparison and should make waste evident.

Flow production: A philosophy that rejects batch, lot, or mass processing as wasteful. Product should move (flow) from operation to operation in the smallest increment, one piece being ideal. Product should be pulled from the preceding operation only as it is needed. Often referred to as "One-Piece Flow," only quality parts are allowed to move to the next operation.

Gage capability study: A method of collecting data to assess the variation in the measurement system and compare it to the total process variation.

Green Belt: An individual trained to assist a Black Belt. This individual may also undertake projects of a lesser scope than Black Belt projects.

Hanedashi: Device or means of automatic unload of the work piece from one operation or process, providing the proper state for the next work piece to be loaded. Automatic unloading and orientation for the next process is essential for a *"Chaku-Chaku"* line.

Heijunka: Production-leveling process. This process attempts to minimize the impact of peaks and valleys in customer demand. It includes level production volume and level production variety.

Hoshin action plan: Form used by the team to detail specific activities required for success, milestones, responsibilities, and due dates.

Hoshin implementation plan: A form used to track performance (plan vs. actual) on Policy Deployment objectives. Usually reviewed with top management on a monthly basis, but reviewed by the Policy Deployment team more frequently.

Hoshin Kanri: A strategic decision-making tool that focuses resources on the critical initiatives to accomplish organizational objectives. This process links major objectives with specific support plans throughout the organization.

Hoshin strategic plan summary: Form used to show relationships between 3- and 5-year objectives, improvement priorities, targets, resources required, and benefits to the organization.

Input: A resource that is consumed, utilized, or added during a process.

Jidoka: Automation with a human touch or transferring human intelligence to a machine. This allows the machine to detect abnormalities or defects and stop the process when they are detected. Also known as Autonomation.

Just-in-Time (JIT): A strategy that concentrates on delivering the right products in the right time at the right place. This strategy exposes waste and makes continuous improvement possible.

Kaikaku: Radical improvement to eliminate waste.

Kaizen: Japanese for "continuous improvement." The term is composed of *kai*, meaning "to take apart" and *zen*, meaning "to make good." Based on the philosophy that what we do today should be better than yesterday and what we do tomorrow should be better than today, never resting or accepting status quo.

Kaizen event: A planned and structured event to improve an aspect of a business.

Kanban: Japanese term meaning "signboard" or "signal." It is a means of communicating a need for products or services. It is generally used to trigger the movement of material where One-Piece Flow cannot be achieved, but is also used to signal upstream processes to produce product for downstream processes.

Keiretsu: A grouping of Japanese companies that allows each to maintain operational independence but also to establish a permanent relationship with other members in the group.

Key Performance Indicator (KPI): A method for tracking or monitoring the progress of existing daily management systems.

Lead time: The total time required to deliver an order to the customer.

Machine cycle time: The time from when the "start" button is pressed until the machine returns to the original starting position.

Manufacturing lead time: The total manufacturing time beginning from raw material to creating the final, saleable product.

Material Requirements Planning (MRP): A computerized system to determine the quantity and timing of material supplies on the basis of a master production schedule, a bill of materials, and current inventories.

Metric: A performance measure that is linked to the goals and objectives of an organization.

Milk run: A supply and/or delivery vehicle that is routed to various locations to pick up or deliver products and supplies. A milk run can be external to customers or suppliers. A milk run can also be internal to a factory with a material handler.

Mission: A statement of an organization's purpose.

Mixed model: A value stream that accommodates multiple product models.

Muda: Japanese for waste.

Multiskilled worker: Associates at any level of the organization who are diverse in skills and training. They provide the organization with flexibility and grow in value over time. Essential for achieving maximum efficiencies for JIT.

Mura: Japanese for unevenness.

Muri: Japanese for unreasonableness.

Noise: Unexplained variability in a response.

Non-value-added (NVA): Those process steps that take time, resources, or space, but do not transform or shape the product or service toward that which is sold to a customer. These are activities that the customer would not be willing to pay for.

Operator cycle time: The time for an operator to complete one cycle of an operation. The total operator cycle time includes walking, loading, unloading, and inspection.

One-Piece Flow: A manufacturing process in which product moves one piece at a time through all necessary operations.

Output: A product or service delivered by a process.

Pareto chart: A vertical bar graph for attribute or categorical data that shows the bars in descending order of significance, ordered from left to right. Helps to focus on the vital few problems rather than the trivial many. An extension of the Pareto Principle suggests that the significant items in a given group normally constitute a relatively small portion of the items in the total group. Conversely, most of the items will be relatively minor in significance (i.e., the 80/20 rule).

Parts per million (PPM): A measure of the number of defective units per million opportunities. A Six Sigma level process is 3.4 defects per million.

Pilot cell: An experimental exercise in a cell to determine the viability of a concept.

Plan-Do-Check-Act (PDCA) cycle: PDCA is a repeatable four-phase implementation strategy for process improvement. PDCA is an important item for control in Policy Deployment. Sometimes referred to as the Deming cycle.

Poka-yoke: A Japanese expression meaning mistake-proof. A method of designing production or administrative processes that will, by their nature, prevent errors. This may involve designing fixtures that will not accept an improperly loaded part.

Policy Deployment: See *Hoshin Kanri*.

Process: An activity that blends inputs to produce a product, provide a service, or perform a task.

Process map: A visual representation of the sequential flow of a process. Used as a tool in problem solving, this technique makes opportunities for improvement apparent.

Production smoothing: See *Heijunka*.

Productivity: Output per unit of input (e.g., output per labor hour).

Pull: A system in which replenishment does not occur until a signal is received from a downstream customer.

Push: Conventional production in which product is pushed through operations on the basis of sales projections or material availability.

Quality characteristic: An aspect of a product that is vital to its ability to perform its intended function.

Queue: Inventory authorized by a push signal.

Return on Investment (ROI): A ratio of profit relative to the money invested. It is a measure of the income provided by an investment.

Rework: An activity to correct defects produced by a process.

Root cause: The ultimate reason for an event or condition.

Sensei: A teacher with a mastery of a body of knowledge.

Setup time: The time between the last good piece to the first good piece of the next product.

***Shusa*:** A strong team leader in the Toyota product development system.

Sigma (σ): Standard deviation of a statistical population.

Sigma capability: A measure of process capability that represents the number of standard deviations between the center of a process and the closest specification limit. See *Sigma level*.

Sigma level: A measure of process capability that represents the number of standard deviations between the center of a process and the closest specification limit. See *Sigma capability*.

Signboard: English for the Japanese term *kanban*. See *kanban*.

Single-Minute Exchange of Dies (SMED): Method of increasing the amount of productive time available for a piece of machinery by minimizing the time needed to change from one product to another. This greatly increases the flexibility of the operation and allows it to respond more quickly to changes in demand. It also has the benefit of allowing an organization to greatly reduce the amount of inventory that it must carry because of the improved response time while maximizing ROI and EVA.

Six Sigma: A quality improvement and business strategy that emphasizes impacting the bottom line by reducing defects, reducing cycle time, and reducing costs. Six Sigma began in the 1980s at Motorola.

Spaghetti chart: A map that illustrates the path of a product as it travels through the value stream.

Standard: A prescribed documented method or process that is sustainable, repeatable, and predictable.

Standard deviation: A measure of variability in a data set. It is the square root of the variance.

Standardization: The system of documenting and updating procedures to make sure everyone knows clearly and simply what is expected of them. Essential for the application of the PDCA cycle.

Standard Work: A tool that defines the interaction of people and their environment when processing a product or service. It details the motion of the operator and the sequence of action. It provides a routine for consistency of an operation and a basis for improvement. Standard Work has three central elements: takt time, Standard Work Sequence, and Standard Work-in Process.

Standard Work-in-Process: The minimum amount of material for a given product that must be in process at any time to insure proper flow of production.

Stretch goal: A goal designed to create out-of-the box thinking for breakthrough improvement.

Supermarket: An inventory storage location authorized by a kanban pull system.

Supplier partnership: An approach to business that involves close cooperation between the supplier and the customer. It provides benefits and responsibilities that each party must recognize and work together to realize.

Takt time: The frequency with which the customer wants a product and how frequently a sold unit must be produced. The number is derived by dividing the available production time in a shift by the customer demand for the shift. Takt time is usually expressed in seconds.

Target cost: The cost a product cannot exceed for the customer to be satisfied with the value of the product and the manufacturer to obtain an acceptable return on investment.

Throughput time: The total time for a product from concept to launch, order to delivery, or raw material to customer delivery.

Total cost: The market value of all resources used to produce a good or service.

Total Productive Maintenance (TPM): Productive maintenance carried out by all employees. It is based on the principle that equipment improvement must involve everyone in the organization, from line operators to top management.

Total revenue: The price of a product multiplied by the quantity sold in a given time period.

Total utility: The amount of satisfaction obtained from the entire consumption of a product.

Toyota Production System (TPS): A manufacturing model built on reducing lot sizes to allow for flexibility, control of production parts, and the logical arrangement of production equipment.

Utility: The pleasure of satisfaction obtained from a good or service.

Value: A capability provided to a customer for an appropriate price.

Value added: Any process or operation that shapes or transforms the product or service into a final form that the customer will purchase.

Value stream: All activities required to design and produce a product from conception to launch, order to delivery, and raw materials to the customer.

Value stream mapping: A process for identifying all activities required to produce a product or product family. This is usually represented pictorially in a value stream map.

Variance: A measure of variability in a data set or population. Variance is equal to the squared value of standard deviation.

Variation: A process is said to exhibit variation or variability if there are changes or differences in the process.

Visual control: Visual regulation of operations, tool placement, etc. that provides a method for understanding a process at a glance.

Visual management: Systems that enable anyone to immediately assess the current status of an operation or given process at a glance, regardless of their knowledge of the process.

Voice of the customer (VOC): Desires and requirements of the customer at all levels, translated into real terms for consideration in the development of new products, services, and daily business conduct.

Waste: Also known as *Muda*. Any process or operation that adds cost or time and does not add value. Eight types of waste have been identified: (1) overproduction, (2) waiting or idle time, (3) unnecessary transportation, (4) inefficient processes, (5) unnecessary stock on hand, (6) motion and efforts (7) producing defective goods, and (8) unused creativity.

Work-in-process: Material in the process of being converted into saleable goods.

Work sequence: The specific order in which an operator performs the manual steps of the process.

Additional Resources

Continuing education is vital. To improve the Lean processes at your company and increase your professional development, I have found the following books quite helpful on the journey to a Lean enterprise.

Bodek, N. *Kaikaku: The Power and Magic of Lean*. PCS Press: Vancouver, WA, 2004.

Brown, M. *Get It, Set It, Move It, Prove It: 60 Ways to Get Real Results in Your Organization*. Productivity Press: New York, 2004.

Hirano, H. *5S for Operators: 5 Pillars of the Visual Workplace (For Your Organization!)* Productivity Press: New York, 1996.

Identify Waste on the Shopfloor. Productivity Press: New York, 2003.

Jackson, T. *Hoshin Kanri for the Lean Enterprise: Developing Competitive Capabilities and Managing Profit*. Productivity Press: New York, 2006.

Keyte, B. & Locher D. *The Complete Lean Enterprise: Value Stream Mapping for Administrative and Office Processes*. Productivity Press: New York, 2004.

Mann, D. *Creating a Lean Culture: Tools to Sustain Lean Conversions*. Productivity Press: New York, 2005.

Martin, K. & Osterling M. *The Kaizen Event Planner*. Productivity Press: New York, 2007.

Rother, M., J. Shook, J. Womak, & D. Jones. *Learning to See*. Lean Enterprise Institute: Cambridge, MA, 2003.

Shinkle, G., R. Gooding, & M. Smith. *Transforming Strategy into Success: How to Implement a Lean Management System*. Productivity Press: New York, 2004.

Wincel, J. *Lean Supply Chain Management: A Handbook for Strategic Procurement*. Productivity Press: New York, 2004.

Womak, J., D. Jones, D., & D. Roos. *The Machine that Changed the World: The Story of Lean Production*. Scribner: New York, 1990.

Index

Pages numbers in italic indicate figures.

Author

Beth Cudney is currently an assistant professor at Missouri University of Science and Technology in Rolla, MO. She teaches Six Sigma, Design for Six Sigma, and Quality Philosophies and Methods.

Beth worked for 7 years in the automotive industry in various roles including Six Sigma Black Belt, Quality/Process Engineer, Quality Auditor, Senior Manufacturing Engineer, and Manufacturing Manager.

Beth received the 2007 ASQ A.V. Feigenbaum Medal and the 2006 SME Outstanding Young Manufacturing Engineering Award. Beth is an ASQ Certified Six Sigma Black Belt, Certified Quality Engineer, Manager of Quality/Operational Excellence, Certified Quality Inspector, and Certified Quality Improvement Associate.

Beth received her Bachelor of Science in Industrial Engineering from North Carolina State University. She received her Master of Engineering in Mechanical Engineering with a Manufacturing Specialization and Master of Business Administration from the University of Hartford, and her Doctorate in Engineering Management from the Missouri University of Science and Technology.

Printed and bound by CPI Group (UK) Ltd, Croydon, CR0 4YY

23/10/2024

01777685-0016